PENGUIN BOOKS

HUMAN MINDS

'*Human Minds* belongs on the same shelf as Daniel Dennett's
Consciousness Explained ... A heartfelt book,
full of interesting insights' – Will Self in the *Independent*

'In this thoughtful and evocative book Margaret Donaldson
shows that academic psychology is not "merely" academic. She
applies fascinating new discoveries on the growing awareness of
the world in babies and children to how adults can discover
themselves, and how to live more effectively and with greater
happiness in the modern world' – Richard L. Gregory

'It blows a breath of fresh air into a discipline that ... sometimes
seems on the verge of premature senility' –
Mihaly Csikszentmihalyi in *The New York Times Book Review*

'She is admirably lucid on the nature of spirituality,
and I doubt whether mystical experience, insofar as it is
describable, has ever been better described' –
Anthony Daniels in the *Sunday Telegraph*

'A stimulating discussion of the way in which we learn
how to think and the implications of the relationship between
thought and emotions ... she is gratifyingly undaunting
to the lay reader' – Penelope Lively in the *Spectator*

'The theory is really impressive ... Donaldson writes
persuasively and shows a deep knowledge of research on
children's intellectual and emotional development' –
Peter Bryant in *The Times Higher Education Supplement*

ABOUT THE AUTHOR

Margaret Donaldson is Emeritus Professor of Developmental Psychology at the University of Edinburgh. Her previous publications include *A Study of Children's Thinking* (1963) and *Children's Minds* (1978). She also writes for children.

MARGARET DONALDSON

———————

HUMAN MINDS

AN EXPLORATION

PENGUIN BOOKS

In memory of my mother
Nan Lennox Donaldson
for the childhood she gave me

—————————————

PENGUIN BOOKS
Published by the Penguin Group
Penguin Books USA Inc., 375 Hudson Street, New York, New York 10014, U.S.A.
Penguin Books Ltd, 27 Wrights Lane, London W8 5TZ, England
Penguin Books Australia Ltd, Ringwood, Victoria, Australia
Penguin Books Canada Ltd, 10 Alcorn Avenue, Toronto, Ontario, Canada M4V 3B2
Penguin Books (N.Z.) Ltd, 182–190 Wairau Road, Auckland 10, New Zealand

Penguin Books Ltd, Registered Offices: Harmondsworth, Middlesex, England

First published in Great Britain by Allen Lane The Penguin Press,
an imprint of Penguin Books Ltd, 1992
First published in the United States of America by Allen Lane The Penguin Press,
an imprint of Viking Penguin, a division of Penguin Books USA Inc., 1993
Published in Penguin Books 1993

1 3 5 7 9 10 8 6 4 2

Grateful acknowledgment is made for permission to reprint an excerpt from
"Burnt Norton" from *Collected Poems: 1909–1962* by T. S. Eliot. Copyright 1963
by Harcourt Brace Jovanovich, Inc.; copyright 1963, 1964 by T. S. Eliot.
Reprinted by permission of the publisher.

ISBN 0-7139-9081-3 (hc.)
ISBN 0 14 01.7033 2 (pbk.)
(CIP data available)

Printed in the United States of America
Set in Monophoto Bembo

Contents

Preface

It is an interesting experience, if at first a disconcerting one, to follow a line of argument starting from familiar ideas and then to find oneself moving in an unexpected direction towards outcomes not foreseen.

The origins of this book lie in some unanswered questions which stayed with me after I had written *Children's Minds*. Its conclusions were to me like signposts pointing beyond themselves, and after a time the urge to find out what was there became compelling. However, when I first tried to do this I believed that I knew what the general lie of the land would be. And I thought that a resting-place in the shape of further conclusions would not be too hard to reach. I turned out to be wrong. The exploration that resulted has taken me a long time and has led me far from the kind of territory which I knew well.

At one stage a perceptive critic who had been kind enough to read a draft wrote as follows, thinking of the journey as a maritime one: 'I feel you are sailing into such dangerous waters that I would be letting you down badly if I didn't stand on the bank waving and shouting and trying to get you to steer another course.' I smiled at the image and at the same time took him seriously. But what other course? There seemed to be no other course open to me except to turn back and abandon the enterprise. This I did not feel able to do.

What chiefly worried my critic was the move that takes place in the middle of the book from psychological to historical discussion. I reached the conclusion, however, that the recourse to both psychology and history was necessary if I was not to dodge some of the most important topics to which the developing arguments had brought me. I knew very well that this move was risky. I felt exposed and vulnerable, as those who cross disciplinary boundaries must always do if they face the extent of their liability to error. However, fear of error ought not to induce paralysis. Also it is worth remembering that we are not immune from mistakes within the disciplines that we call our own. It is even arguable that these mistakes are the more dangerous kind since they bear the stamp of some sort of authority. This by itself means that they tend to be more deeply rooted. They are usually entangled with our self-esteem and it is hard to pull them out.

I did, of course, take historical advice, and I was fortunate enough to have Rosalind Mitchison's scholarly comments available to me. I hope I used them well. I certainly enjoyed my talks with her and gained much from them. I am most grateful for her guidance.

As the arguments in the book evolved it was not only in the direction of history that I had to cross boundaries. It also came to seem necessary to write about certain aspects of religion – and now I was even more acutely aware of needing help. I received this generously from John McIntyre. To have the benefit of his wide understanding of theology, of philosophy and indeed of psychology was a reassurance and a privilege.

In the specific matter of Buddhism I had further assistance from two people with particular knowledge of that religion: Rod Burstall and Tom Thorpe. Both made comments which I found illuminating, and which certainly enabled me to improve what I had written. I do still have to acknowledge that my discussion of Buddhism is inescapably that of an outsider with an unusual point of view. But I write with respect, and with appreciation of what I have personally gained from my studies of Buddhist teachings and from the guidance of these two friends.

I also needed, and was generously given, help from within my own discipline. I am grateful to Jerome Bruner for important advice, which I took as far as I felt able. But some of his thoughts raised major questions that went beyond my scope here. Perhaps they will come to serve as a new set of signposts.

Jess Reid, Alison Elliot and Morag Donaldson kept up their support over a period of years, patiently reading one section after another with remarkable fortitude, writing criticisms and discussing ideas with me at length. I owe a great debt to them. Valuable support of a different kind, also sustained over a long time, came from Shimon Abramovici and Tony Fallone, who greatly assisted me with library research.

As the work came nearer to a conclusion I was fortunate to have comments from David Bloor, Robin Campbell, Kath Davies, Tony Fallone, Barbara Gardiner, Robert Grieve, Paul Harris, David Hay, Lynne Murray, Chris Pratt, Martin Hughes, Janet Simpson, Jane Turnbull, Colwyn Trevarthen and Jennifer Wishart. They showed me many possibilities of improvement and I am grateful to them all.

I want also to thank my agent, Caradoc King, whose enthusiasm for the book was a great encouragement to me, and my editors, Jon Riley in London and Dawn Seferian in New York, for important suggestions and much considerate support.

Marcia Wright, Avril Davies and Rosemary Teacher typed most of the manuscript, and it was often not easy to decipher. I appreciate their patience, skill and care.

It is a convention that one's spouse should be thanked last. My husband, Stephen Salter, has certainly earned this distinction. He is the only person, so far as I know, who has gone through the whole book from back to front. This is not as absurd as it sounds. He had offered to put a large number of revisions on to a disk for me and he started at the end so that the insertion of one change would not affect the page number of the next. This made the task easier but still it was unavoidably boring and long. I am very grateful to him for undertaking this considerable chore.

Beyond this, however, I thank him for a more sustained contribution: his steady support for the whole enterprise, even when the goal seemed to be receding instead of drawing nearer. He is a man who understands absorption and commitment and who knows well that, while ideas can appear in a flash, the working out of their implications may take a very long time.

Edinburgh
January 1992

PART ONE

DEVELOPMENT IN CHILDHOOD

I

Modes of Mind: An Introduction

Some years ago a man in the United States managed to put one hundred rattlesnakes into a bag in twenty-eight seconds. It was a record at the time. Most likely it has been surpassed already. The man's behaviour was odd, deviant even. Most of us do not go around putting rattlesnakes into bags for the sole purpose of showing how good we are at doing it. Yet this strange act reveals something fundamental about human nature. In spite of its oddity, it illustrates an attribute that we all share.

Many suggestions have been made about the qualities that mark us out as a species. Among the candidates have been ability to use language, skill in opposing thumb and forefinger for the highly controlled use of a tool, knowledge of good and evil – and so on in some variety, so that there is clearly no point in trying to find *the* one. But the man with the rattlesnakes demonstrates a characteristic that is crucial for an understanding of the human species: he illustrates our remarkable capacity for forming novel kinds of purpose. One of the most striking things about us is that we are highly prolific 'intention generators'. We set goals for ourselves of the most diverse kinds. One person forms the purpose of driving people into gas ovens. Another forms the purpose of rescuing abandoned children from the streets of Calcutta. Both are human beings.

However, this diversity does not mean that we have no goals in common. There are some kinds of general purpose that are extremely pervasive among us – universal indeed, except in certain pathological conditions – such as the goal of understanding or making sense of things and the goal of communicating with one another.

Also, of course, we share with the other animals certain physiological urges which we experience as hunger, sexual desire, and so on. However, it is characteristic of us that we are capable of transcending these urges, though not easily. We may set ourselves some new goal which requires that we deny them. We may decide to go on hunger strike or take lifelong vows of chastity.

I am not concerned here with the philosophical problem of free will but with the psychological fact of the experience of choice. In so far as we have this experience, we *take* responsibility. We commit ourselves. In many cases this commitment does not last. But it may occasionally happen that, from all the possibilities open to us, we come at some stage to a choice that proves to be definitive and exclusive, so that thereafter, to adapt Emily Dickinson's words very slightly, we close the valves of our attention like stone.[1]

The human beings who settle in this way on a single all-absorbing goal are few in number, and they tend to belong to certain cultural groups which favour this kind of commitment. But their existence reveals already that as a species we possess another attribute beyond that of being able to generate intentions that are new and varied. We possess also the ability to pursue our goals with great tenacity. This tenacity has a number of sources, but prominent among them is the fact that our purposes are apt to be accompanied by very powerful feelings. Thus they become important to us; and in the extreme case they can become more important than life itself.

It is not an everyday matter for someone to die rather than abandon a purpose. Yet such an event is common enough for us not to be specially surprised when we hear about it; and we are apt to feel admiration and respect even when we consider the behaviour misguided, the goal not worth dying for.

In spite of these facts of experience and observation, a number of serious attempts have been made to account for human behaviour without having recourse to the notion of intention or purpose at all. The notion, however, is one that tends to reappear in some guise or other within psychology, no matter how hard one tries to keep it out. And it is ironic that the attempt to keep it out is generally itself sustained by a passionate aim: the aim of being 'scientific' in a manner modelled upon the activity of the physicists.

We should not, on the other hand, naively suppose that what people do is always and only determined by what they believe themselves to be trying to do. There is much of which we are unaware. Also we may deceive ourselves. These are topics to which later chapters will return.

The devising of novel purposes comes readily to us because we have brains that are good at thinking of possible future states – at considering not merely what is but what might be.

We exist in a world of 'hard fact', but we can imagine it as changed; and from a very early age we know that, within certain limits, we are able to change it. It matters very much to us to find out how these limits are set, an activity closely related to the general purpose of understanding what the world is like. In this context 'the world' includes other people as a most important component. And, if we have any wisdom, it also includes ourselves.

Human thought deals with how things are, or at least with how they seem to us to be, but it does this in ways that typically entail some sense of how they are not – or not yet. It deals with actuality and with possibility; but some recognition of possibility is already entailed even in the discovery of actuality whenever this is achieved by the characteristically human means of asking questions. Is it like this? Or is it perhaps like that?

We all ask many questions of this kind when we are young, and some of us retain the habit. A few even become fascinated by a single question and devote their lives to finding an answer. When Einstein was sixteen he asked himself a question that has since become famous: 'If I were to travel with a ray of light, what would I see?' He was inviting himself to consider the possibility, never realized in human experience, of travelling with light at its own speed.[2]

How he came to appreciate the significance of just this question, and so to make it *his* question, is a matter of great interest. But for now the point is that, having chosen it, he spent the next ten years searching for an answer. That answer, when it was finally found, took the form of the special theory of relativity, which led to a major revolution in ideas about the nature of space and time.

Years later Einstein said of himself: 'I have no special gift. I am only passionately curious.'[3]

We may dispute the first part of this statement, but as to the second we must take his word – and without surprise. The tenacity needed to get his answer had to come from some strong emotion. Without that he surely would have given up. On the other hand, he could not have reached that answer if he had not been adept at a certain kind of thought – a kind that has given us immense power and, with the power, many problems. We shall never understand the position of humanity today unless we understand not only the nature of this thinking but also how it relates to the rest of mental functioning, which in turn depends on an understanding of how the mind develops from birth on. Knowledge of being rests upon knowledge of becoming, in this case certainly, if not always.

In many ways different minds develop differently. This is an obvious truth, and it is implied in much that has been said already about the generation of goals and the experience of choice. However, there is also commonality. There is a framework within which the deviations are contained.

My account of the common framework entails the distinguishing of four main modes of mental functioning. These come in succession upon the scene as we grow older, but they do not replace one another. None of them is ever lost, except in severe injury or illness. But within each mode change occurs over time. They are not static. For instance, the functioning of the first mode in infancy is a very different matter from the functioning of that same mode in adulthood.

In defining the modes two kinds of criterion are used. First there is the *locus of concern*.

What I mean here by 'concern' is best captured by saying that a mind's concern at any given time is what its percepts, thoughts, emotions or actions are *about*. If they were not about anything, there would be no concern, and hence, of course, no locus of concern, which would mean that none of the modes, as here defined, would be in operation.

Being about something is often regarded as a fundamental attribute of mental states or acts. And this quality of 'aboutness' is known in philosophy by the technical term 'intentionality'.[4] I have preferred to use the word 'concern' because of the risk that 'intentionality' will be taken as a property of intentions only, whereas ideas, hopes, beliefs and the like are also included in the technical meaning, as are signs and symbols that stand for something else. 'Intentionality' in this philosophical sense goes back to the Scholastic philosophers and their use of the Latin word '*intendere*' meaning 'to stretch or reach towards'.

We shall speak, then, of loci of concern. Four of these will be distinguished; and they will serve to specify the four main modes. The other kind of criterion, to which we shall come shortly, yields subdivisions of the four main categories. (When these subdivisions are referred to they will still be called 'modes' rather than 'submodes' to avoid cumbersome terminology.)

In the first mode – the only one available to the young infant – the locus of concern is always the present moment, the directly apprehensible bit of space, the 'here and now'. This mode is called the *point mode*.

Later other loci become possible. For example, the second mode, which is called the *line mode*, has a locus of concern that includes the personal past and the personal future.[5] When we function in the line mode we look back to what has taken place in our lives so far, or forward to what we can consider as possible happenings. In due course the scope is extended beyond the range of personal experience, but by definition concern is still with specific events, actual or conceivable.

These two examples should already make it clear that 'locus of concern' is defined in terms of space-time. Notice, however, that *locus* of concern is not the same as *focus* of attention. Within the here and now – and still more obviously within the personal past and future – there is a wide range of things that the mind may pick out as worthy of special interest and consideration.

The second kind of criterion, needed to define the subdivisions, is the manner in which the components of experience – perceiving, thinking, and so on – are linked or separated. In the earliest manifestations of the point mode these are inextricably intertwined. Then, step by step, certain separations become possible; and these new possibilities bring with them dramatic change in what the mind can do.

Development ranges outwards in two directions from the initial tight compactness of the first mode. In one direction there is extension in what is possible by way of locus of concern. In the other there is separation of the different components of mental life so that they can function with an increasing measure of independence.[6]

But what are these components? In defining the modes, four are used: perception, thought (in the sense of knowing, understanding, solving problems), emotion and action (that is, directed action as distinct from reflex movement and from movement that is passive, as when we are caused to move by something or someone else).

It is at once evident that a different set of components could have been chosen. What about imagination, for instance, or memory, or language? I can only answer that my choice of criteria reflects my judgement about the most significant distinctions – those that yield the best insight into how the mind develops. However, the contributions of such activities as imagining, remembering and using language will repeatedly enter into the discussions that follow.

Of the four defining components, one calls for further comment now: we must be clear at the outset about what is meant here by emotion. In particular how do 'emotions' differ, if at all, from 'feelings'? The crux is that emotions are our value feelings. They mark importance. We experience emotion only in regard to *that which matters*.

However, it is necessary now to draw a distinction, for more than a single kind of response to importance is open to us. Suppose that one is looking at a fine painting, a great work of art. There are then two ways of recognizing its quality. On the one hand it is possible to assess it as the expert does, using various bits of evidence to decide whether it was really painted by the supposed artist, whether it is in good condition, what it might fetch at auction, and so on. This is a value *judgement* and it may be performed quite unemotionally. But on the other hand we may stand before the painting doing none of these things and directly *feel* its significance. We may be 'moved'. For instance, I cannot look at the Rembrandt self-portrait in the Scottish National Gallery without encountering value directly, felt in the body. The experience is profoundly emotional. The painting matters to me.

The difference between value judgements and value feelings is evidently great and the distinction needs to be noted. In practice, though, judgements and emotions are not always − or, I think, usually − so distinct as in the above example. Commonly they are interconnected, even interdependent. Thus when emotion arises the experience very often contains a powerful component of judgement or interpretation. For example, imagine yourself seeing a point of light in darkness. Does it give rise to emotion and, if so, what emotion? That will depend.

Consider two conditions. In the first you are lying on a mountainside with severe injuries, so that you cannot move. A friend has gone for help. In the circumstances, you will take the light to *mean* rescue. It will have high personal importance and you will experience strong emotion: relief, hope, joy. You will feel this powerfully with your body.

For the second condition suppose that you are a fugitive, hiding in a wood. You know that people are searching for you with hostile aims. In this case you will take the light to mean danger. Once again the importance for you will be great and the emotion correspondingly strong − but different. You will experience fear. The extent of this fear will be affected also by your assessment of your ability to cope with the situation, as Richard Lazarus has pointed out.[7]

However, it is easy to imagine less dramatic circumstances in which the point of light occasions no emotion at all – circumstances where its presence or absence does not matter, where it is, as we say, of no significance. An event that is of no significance gives rise to no emotion.

The word 'significance' must not be narrowly interpreted here. Significance, for human beings, does not arise only in connection with such things as physical threat or rescue from danger. The range is vastly wider. For instance, we may love someone. The loved person is one who matters; and the existence of a significant other is then the source of extremely powerful and varied feelings. Also, as we have seen, art may be perceived as having deep significance and will then generate corresponding emotion. The same applies to music.

The relations between value judgements and value feelings can get more complex than we have so far seen. Consider the following case. A friend of mine tells me that he cannot listen to 'God Save the Queen' without a surge of strong emotion. But he is embarrassed by this because he judges that the monarchy is anachronistic. Does this, then, cast doubt on the claim that his involuntary response to the national anthem has to do with what matters to him? I think not. What is happening is that an old source of value, now disowned but still rising up as feeling, is coming into conflict with a later judgement. This in turn gives rise to a new value feeling: the embarrassment. And clearly the source of importance from which the embarrassment stems is the image one has of one's self. The integrity of that image is threatened if one experiences – in spite of oneself, as we say – an emotion of which one cannot approve. It is not an uncommon event.[8]

It is evident that, on this analysis, we have many feelings which are not emotions. Take, for instance, hunger or pain. These are not emotions, though they may be accompanied by emotion. We may be angry at the cause of the pain or, on the other hand, we may know that the pain is being produced by an attempt to help us. We may be afraid that more and worse pain will follow to no foreseeable limit. Or else we may know that the pain is brief and limited, soon to be over. And there is even the condition called masochism where pain of certain kinds is, for complex reasons, sought and in some sense enjoyed. Thus pain, however strongly felt, is no more − or not much more − *inherently* emotive than the spot of light. So, too, with hunger. We might have decided to fast.

Emotions are a subclass of feelings. As with many other categories, this one has core or prototypical examples and becomes a little fuzzy towards the edges. That is, one might sometimes reasonably be unsure whether to call a feeling an emotion or not.

For instance, is boredom an emotion? It is no more necessary to have a firm answer than to be able to determine whether a blue-green colour is blue or green, or to be able to say precisely where to draw a borderline between friendship and acquaintance. The important thing to recognize is that a prototypical emotion − a clear case − will be a strong indication that something matters to the person who feels it.

If we return now to a consideration of the nature of the modes, it will be evident that the two criteria (locus of concern and separation of components) are both implicated in the movement from the point mode to the line mode. When the locus of concern shifts from the present moment into the past or the future, this shift entails the 'separating out' of perception and action from the other − still interwoven − functions, so that mental life in the second mode goes on without them. For we cannot directly perceive and act on what is not here and now.[9]

No attempt will be made now to describe the later modes in detail, for an account of them would be obscure in advance of the discussions that are to follow. The modal structure was not devised a priori, but emerged gradually and through many modifications from a consideration of the evidence and arguments presented in the next seven chapters. The best way to understand it may be to follow the tale as it unfolds. However, some people may feel the need for a more complete overview at the start. Or an overview may prove helpful at a later point if something is unclear. So the bare bones of the structure are set out in an appendix on p. 267.

Meanwhile, however, one or two general points can be made that will give some idea of the nature of the developments we shall have to consider.

Beyond the line mode the major step that is taken consists in movements towards the impersonal. That is, the mind starts to be able to function in ways that achieve some independence from personal goals. For instance, it becomes possible to think about problems of some generality.

This entails movement away from concern with specific events in specific lives. It also entails movement in the direction of increasing separation of mental functions one from the other. Just as thoughts and emotions become detachable from perception and action with the advent of the second mode, so now thoughts and emotions become detachable from certain personal purposes, and to some extent from one another.

I am not talking here about the state of mind in which thought is allowed to drift without conscious purpose: the state we sometimes call day-dreaming. This is an interesting state but it is not what is involved in the later modes as here defined. These entail a different contrast. On the one hand, there is thought directed to the solution of some personal problem – how to achieve or avoid some happening – and this is typical of the second mode. On the other hand, it is also possible for human beings to think with the aim of understanding some aspect of the way things are. In this sense there is movement towards impersonality, though the movement may be powered by intense personal curiosity.

The notion of *thinking* of this kind is familiar enough. But what about emotion? Can we take steps towards impersonality in respect of our emotions also? And, if so, what kind of experience then ensues?

Most people in the kind of culture that we call 'Western' will probably think this an odd idea at first. It is, however, one with which the second part of the book, from chapter 9 onwards, will be much concerned.

The process of 'opening out' in those two directions is the one that I have previously called *disembedding*.[10] The notion is that, to begin with, the mind functions in the context of its own totality and in the external context of people, things and happenings. Each aspect of functioning is supported by inclusion within this whole so that mental life is embedded in a matrix – a matrix that is at once helpful and limiting. Then step by step the unity breaks up and new ways of doing and experiencing become possible. These are much more difficult to initiate – and especially to sustain – but they offer great gains in flexibility and in power.

Previously I spoke of embedding and disembedding as if there were only a simple dichotomy to recognize. Finer distinctions, however, will now be drawn.

2

Some Human Ways of Dealing with Hard Fact

'Go, go, go, said the bird: human kind
Cannot bear very much reality.'

– T. S. Eliot

Human beings know more about life's dangers than other animals, and have to deal with a greater range of fears. Because we judge ourselves and know ourselves to be capable of acting foolishly or disgracefully, we fear shame, guilt and loss of self-respect. We know the risks of accident or illness and the certainty, if we avoid these, of loss of strength in old age. We know that thieves break in, that rapists and muggers strike, that jobs can be lost. We know that love can be lost. We know that, sooner or later, we must die.

So is it the case that human beings are typically timid and fearful, burdened as they are with so great a load? Not at all. Compared with other animals we are, as a species, extraordinarily bold – reckless even. Consider how we deal with fire. Other animals flee from it, but long ago we took it into our homes, making it serve our purposes. And if more proof is needed, contemplate any busy motorway.

How, given our knowledge of danger and our wide experience of fear, is this boldness of ours to be explained? It seems sensible to tackle the question by looking first at characteristics of the ways in which we know anything whatsoever. Knowledge of danger is not – could not be – unique. It must share attributes with other kinds of knowing. So are there features of human knowing in general which may help us to understand how it is that unusually extensive knowledge of fear can go along with unusual boldness?

One common but mostly unexamined way of talking about

knowledge is as a *thing* which we receive – an abstract kind of thing, certainly, but having none the less the thing-like property of being able to be handed over. We often speak of 'getting' knowledge as we might talk of getting a refrigerator or a new car – or perhaps of getting praise from someone. Only a manner of speaking? So it is. But it carries the implication, no less powerful for going often unnoticed, that the knower is passive and that knowledge comes to us ready-made. It may also carry the further implication that the receipt of the knowledge leaves us essentially unchanged: we only 'have' a new possession. The picture thus drawn does not correspond to the way we are.

Most of the knowledge that matters to us – the knowledge that constitutes our conception of the world, of other people and of ourselves – is not developed in a passive way. We come to know through processes of active interpretation and integration. We ask questions, which may or may not be put into words and which may or may not be addressed to other people. We have strategies of many kinds for finding out. We struggle – and it can be a long, hard struggle – to make sense.

The notion that we make our own systems of knowledge has been strongly advocated by Jean Piaget. According to him the development of the mind consists largely in the building of 'cognitive structures' and the rebuilding of these structures in new and better-integrated forms over long periods of time.[1]

Others, like Erich Fromm, seem to want to go still further in rejecting the emphasis on knowledge as 'thing-like' (which Fromm would say is dangerously common in our education and in our whole culture) and argue that the kinds of 'coming to know' that matter are better thought of not as changes in what we *have* but as changes in what we *are*.[2]

However, the difference between Fromm and Piaget in this respect is not as great as it might at first seem, since in Piaget's view these 'structures', by their presence or absence, do determine what we are as knowing and thinking beings. They are the stuff of which intelligent minds are made.

Clearly, there is some knowledge, like knowing the year of

Napoleon's birth or the capital of Chile, that leaves us essentially unchanged, especially if we 'get' it simply by being told it. However, there are other kinds of knowledge that alter us profoundly. And it is even more important to recognize that the *processes* of coming to know transform us. This is particularly so when these entail sustained, self-directed effort.

When we discuss the development of the human mind we are talking about processes of self-transformation: processes by which we turn ourselves into different beings. However, in stressing *self*-transformation we should never forget that this is not a solitary effort. We are dependent in the most crucial ways on the help of others. And others may hinder or constrain us also. This is true from early infancy onwards.

A question now arises: to the extent that the processes of coming to know are self-directed, can we choose what we will know? And, by the same token, can we choose what we will not know? Can we refuse to know what we do not want to know?

This is a vast question having many ramifications. There is, first of all, an obvious sense in which we may consciously refuse to undertake the effort that coming to know often entails. James Thurber is said to have remarked: 'I do not understand electricity and I do not want to have it explained.' In a similarly deliberate way children commonly enough refuse to know things that adults try to teach them. There is the story of the boy who advised his young brother not to get his sums right, 'because if you do they'll just keep on giving you more and more of them'.

The matter, however, is not always so straightforward. It is complicated by the fact that we may know in a variety of ways characterized by differing degrees of awareness.

Discussion of this topic makes the use of visual metaphors – terms like *light* and *dark* – almost unavoidable, for they seem to fit the case so well. Some kinds of knowledge are in the light of full awareness. Others are in the shadows, on the edge of the bright circle. Still others are in the darkness beyond.

Let us apply this notion to our everyday knowledge of danger.

As we drive fast along a busy motorway we know that we are doing a dangerous thing, but the fully lit, central parts of our consciousness are not occupied by this knowing. A sudden event, however, can very rapidly bring knowledge of danger 'into the open'. People who witness a bad accident generally drive warily afterwards, at least for a while; and the effect is more pronounced, longer-lasting, for those actually in the crash and harmed by it. Yet before the accident they *knew* perfectly well, in one sense, about the danger.[3]

The same is true of other kinds of hazard. Because we have a considerable ability to manipulate our own consciousness, we generally contrive not to think of them unless circumstances force them upon our notice. This is why a major nuclear accident, like the one at Chernobyl in 1986, had such a marked effect on public opinion, at least for a time. But people differ in their ability to push unpleasant fears out of sight and hold them there.[4]

Knowledge of the kind we have been considering – knowledge that is on the fringe of consciousness – is always ready to move to the centre. It is accessible to us, even if we don't attend to it. We are able to talk about it. Even as we ignore it, we really know it is there.

However, we may also have knowledge which, to continue the metaphor, lies in darkness so deep that we are not aware it is there at all. Thus, to put it paradoxically, we can know without knowing that we know. This at once raises the possibility that we can refuse to know without being aware of having done so.

Consider first an example that refers to a pathological state but is most revealing. When people with normal eyes undergo operations to remove the visual cortex at the back of the brain, they report that they can no longer see. Their conscious experience is of being quite blind. Yet research by Weiskrantz has shown that some visual information is still being received.[5] Weiskrantz encouraged people in this condition to guess what their eyes were looking at. As it turned out, these 'guesses' were so often correct that clearly they were not just guesses at all. Success was too frequent to be attributed to chance. The people had some genuine knowledge of what was before their eyes, even though they believed they saw nothing. They knew without knowing that they knew.

We may safely take it that these people were not refusing to know that they knew: they were unable. The point for the moment is simply to establish that the distinction between *knowing* and *knowing that one knows* can validly be made. It is relevant to the human condition. We are so constituted that it is possible for us to have knowledge of which we are not aware.

Non-pathological examples can readily be found. Language is a rich source, though here a different formulation is appropriate. It is not so much that we don't know *that* we know. Rather we don't know precisely *what* we know. When we use our native tongue we operate with implicit knowledge most of the time, and we often have great difficulty if we try to make it explicit. My own first awareness of this came when I went to France as a student and earned my living by taking English conversation classes with senior pupils at the local *lycée*. The trouble was that sometimes they expected me not just to talk but to explain the rules. To realize how hard this can be, think about the words *some* and *any*. If you are a native English speaker, you know just when and how to use these words. The knowledge is so firm and sure that its application requires no thought at all. But try now to say just what it is you know and you will find that it is not at all easy. The knowledge that you built up as you learned the language is not readily accessible to the part of the brain that could reflect on it and put it into words. You do not know in any fully conscious way.

Here again there is obviously no justification for talking of a refusal to know. Rather most of us neglect to develop explicit knowledge of our own language because we do not need it. Linguists are the exception. In choosing language as an object of study they commit themselves to the construction of knowledge systems of an explicit kind. They commit themselves to getting their knowledge into a form that enables them to talk *about* it and not merely *with* it.

The distinction between implicit and explicit knowledge was made in Middle English with the help of a noun which, unfortunately, has been lost from ordinary usage – the noun 'acknowledge'. This means knowledge of the explicit kind. It was substantially different in meaning from the modern 'acknowledgement', though the link is easy to see. There was also a verb 'to acknow' meaning 'to come to know or recognize' or 'to admit or show one's knowledge'. In the latter sense it is quite close to the modern verb 'to acknowledge', but the notion of admission was once only part of the meaning.[6] Both old terms seem valuable and I shall use them.

It is clear, then, that our knowledge is not all of a piece. In some cases we can readily say what we know – say it to others or to ourselves. We can discuss our knowledge and reflect on it. This is *acknowledge*. It is spelled out, available for scrutiny. But what we know is not always ready for inspection. Some of it is kept dark.

Given that it is possible for knowledge to be kept dark, the next question is whether it might ever be kept dark deliberately. Would we ever have good reason to do this on purpose, not just by default or neglect? Are there things that we might deem it best never to acknow? Or are there occasions when we might decide to thrust back into the dark what once had been acknown? What would it profit us?

Human kind can find reality hard to bear. Knowledge is not always comfortable or comforting. When reality, as we experience it, does not suit us, turns out not to conform to our purposes, is not shaped close to the heart's desire, our first impulse is always to change it. We are great manipulators, in the literal sense. We use our hands, together with our brains, to change the inanimate physical world. Also, in the metaphorical sense, we frequently try to manipulate other people, to get them to behave in ways that suit us. But what if reality proves resistant, obdurate? Suppose, for instance, that we come upon some stubborn incompatibility in the nature of existence so that two outcomes, both desired, preclude one another. What then?

Well, then another recourse is open to us. If we cannot change the world, we can perhaps change our conception of it. After all, we construct this conception, we do not just receive it. So we have a good *say* in what it becomes. (Notice this manner of speaking. It is revealing.)

Chapter 5 will discuss the developmental origins of our ability to manipulate our own awareness. There are certainly limits to what we can manage in this regard. Sometimes, the construction collapses. But we are clever architects and masons all the same.

So far I have spoken as if reality – the reality to be changed or, failing that, to be conceived of differently – is external to us. But for human beings the reality that has to be dealt with includes the self. We are self-aware. We know our selves, we experience our selves. And we judge our selves. We like, or do not like, what we know.

Not liking one's self can be a particularly unpleasant experience. When it occurs, there is the usual pair of possibilities (apart from acceptance of the unpleasant truth): to change the self or to change the way we 'see' the self, to change what we are or to change what we acknow. Frequently the second of these must appear as the more attractive option.

Let us now draw together some of the main relevant points and see what they imply.

We come to know things actively. We interpret the world, we struggle to *make* sense. Our most important kinds of knowing do not come ready-made. The world that we interpret includes our selves. We develop concepts or images of our selves. We judge our selves.

We can know in different ways. Some of our knowledge is explicit, out in the open. We know that we know it. We can give an account of what we know, and sometimes of how we came to know it or how we would justify the claim that it is 'knowledge'. But we also have knowledge that is to varying degrees implicit, in the dark – not spoken of, sometimes not able to be spoken of.

We like to 'take control'. Typically, we try to manipulate what we encounter. We try to change the world to suit our purposes.

These purposes are often powerful and passionate. And reality often resists them.

In much of human functioning, knowing and thinking are interfused with emotion. This means that they may be interfused with happiness but also, not infrequently, with distress and pain.

These considerations, taken together, make it highly predictable that we should try to exercise control over aspects of our own knowledge – that we should aim to protect ourselves from unpleasant experience by manipulating our own consciousness. They imply that it might seem to us desirable in certain circumstances not to acknow what we know. In this context, what we know should be taken to include what we remember as having happened, for memories when they arise are often loaded with emotion. They are potent sources of joy or sorrow.

It will be apparent to anyone at all familiar with the writings of Freud that the preceding argument has led us to an idea that is central to psychoanalytic theory. From a consideration of certain general characteristics of human beings and without relying on clinical evidence we have arrived at a notion very close to the Freudian concept of repression or defence: the concept of protecting the ego by manipulating the contents of consciousness.[7]

The Freudian conception of this kind of defence amounts to the claim that we have at our disposal ways ('defence mechanisms' in Freud's terminology) of refusing acknowledge. We can know, yet decline to acknow. I believe this to be a notion of great importance, indispensable if we are to understand ourselves. It is a truth about the way we are.

However, I do not think it helpful in this connection to invoke the Freudian structures of ego, id, and so on. As I see it, what the defence mechanisms mainly serve to defend is the life-space of the present, the only time in which we can do and be. We are fighting for scope – and some measure of peace – in the here and now.

I share Freud's views to the extent of believing that defensive processes begin early in life; but I agree with Daniel Stern that the earliest experiences are not reality-distorting wishes and fantasies but are rather direct attempts to cope with reality. As Stern says, evidence from the study of infants (of a kind not available to Freud) gives us no reason to suppose that the pleasure principle comes first and the reality principle follows.[8] On the contrary, as we shall see later, it now appears that distortions of reality are most likely to arise when later developments of the mind have made possible the manipulation of one's own consciousness. This, however, is still quite compatible with the view that many adult defences are as unsuccessful as they are because of having taken origin in the efforts of an immature mind. For the needful developments that enable one to manipulate one's awareness are not long delayed.

In talking of the defences Freud uses one image which I find illuminating. He likens the activities of a mind shaping its own consciousness to those of an editor revising a text, working towards an acceptable final draft.[9] The various mechanisms then have different editorial counterparts. For example, amnesic repression is equivalent to complete removal of parts of the text, as when a censored newspaper goes to press with blanks where bits have been deleted. Thus whole chunks of a life can be forgotten.

Likewise denial is equivalent to the insertion of 'not':

> not
> 'I do   hate my father.'

Projection is equivalent to changing the subject of a sentence:

> He is
> 'I am evil, lazy, useless.'

Displacement amounts to changing the sentence object:

> enemy
> 'I hate my father.'

And so on. In this way, we write for ourselves an authorised version of our lives. It can truly amount in the end to a Holy Scripture, to be lived by, to be revered. It is then hardly surprising if it is resistant to change, even when it is doing us harm. And it often does us harm to base our lives on such distortions of the truth. For one thing, the version that is 'edited out' does not vanish for ever. Somewhere in the depths it still remains to trouble us. As Freud says, the repressed tends to return. Unbearable reality is not truly rendered bearable in this way.

The preceding argument depends on the notion that we construct our conception of things, including ourselves. It depends on the claim that our experience of the world is experience of an interpretation. As usual, however, it is important not to exaggerate. If this idea is taken too far, there is a danger of stressing subjective experience to such an extent that what happens *to* us, or in any way independently of us, is ignored.

There are several correctives to this. One is to bear in mind that we may *mis*interpret. Reality may not be directly knowable, but to talk of misinterpreting, misunderstanding, mishearing is not nonsense. Interpretations may be better or worse. They may be closer to the way things are or further from it, though which of these is the case may not always be easy to tell. Nevertheless we have certain strategies for deciding. We all know what it is to discover that we have been mistaken.

To take a trivial but clear example, we know that we are subject to certain perceptual illusions. For example, two lines in a figure may appear to be of different lengths. However, we can measure them. If the measuring shows them to be the same length we then know that the perceptual impression, even if it seems compelling, is wrong.

There is satisfaction in dispelling an illusion, unless it is one that serves an important purpose in our lives – and sometimes, fortunately, even then. But why? The fact is that our fondness for shaping things according to the heart's desire is balanced by another more austere aim: the aim of understanding, of getting at the truth. This also is native to us, though our commitment to it varies, both between individuals and within individuals according to precise circumstance.

The second corrective is to consider shared experience. Think, for instance, of the case where two friends go together to a concert. In one sense, although they are together, they do not have the same experience. To stress this is to stress interpretation. What they *make* of the music, as we say, will depend on musical knowledge, sensitivity, and so on. The music will mean something different to each of them. Yet, in another sense that must not be neglected either, they do have the same experience: one set of players sits before them in the concert hall making one set of sounds. And afterwards the two friends can talk to one another about features of their shared experience. They may disagree even as to matters of fact. Did the pianist play a wrong note? But they can discuss this. They have a basis for an opinion, unlike a third friend who stayed at home.

Finally, the third corrective is to think of cases where things happen to us that depend little on interpretation. For instance, we may be hit on the head. Reality then impinges forcefully.

Afterwards, assuming that we survive, we are likely to revert to our usual kind of activity and try to construe the event so as to make it fit in some way or other into what we see as the pattern of our life. We may say: 'Just like me – I'm always the unlucky one.' Or: 'That was God's way of knocking some sense into me, making me go in a new direction.' Or: 'Just think – if that hadn't happened, I'd probably never have met you.'[10]

These construings will affect the memory of what happened. But certain essentials will remain: we *were* hit on the head. Usually we distinguish very well the difference between an event that occurred and one that we imagined. For example, a real rape is a very different matter from a fantasy rape. Yet Freud, who at first believed his patients when they spoke of rape and seduction in childhood (as they often did), came to say later that many of them were not remembering real events, they were recalling early fantasies and could not tell the difference.

It is true that, if a memory/fantasy goes back far enough into childhood, what really happened may not always be easy to establish. In certain circumstances young children seem unsure of the distinction between what they have imagined and what is real. Thus they may fear imaginary ghosts and monsters.[11]

One of the pupils at Bruno Bettelheim's famous Orthogenic School for very seriously disturbed children was a boy who had paranoid fantasies. It turned out that at the age of three he had undergone the dreadful experience of being hanged – but cut down just in time – by his older brother and the brother's friends and had then been terrified by them into telling no one. The boy who suffered this trauma had no conscious memory of the event. In the end, however, the brother told the true story of what had happened.[12]

This is a case of something that appeared in the guise of fantasy but turned out to be based on an actual event. The arguments of this chapter and of the next imply that distortions in the other direction can certainly also occur. However, Freud finally went so far as to claim not just that confusions and distortions arise but that the distinction between happenings and fantasies does not matter: 'up to the present we have not succeeded in pointing to any difference in the consequences whether phantasy or reality has had the greater share in these events of childhood'.[13]

Jeffrey Masson, in the course of an analysis of why Freud shifted his position so dramatically – an analysis which, if well based, does Freud little credit – argues that 'in actuality there is an essential difference between the effects of an act that took place and one that was imagined'.[14] There is indeed.

3

The Point Mode and Its Origins in Infancy

The first mode, which is called the *point mode*, has already been defined as a way of functioning in which the locus of concern is the directly experienced chunk of space-time that one currently inhabits: the *here and now*. Once more, as in the attempt to define emotion, the boundaries are fuzzy. Where does 'now' begin or end? Or again, is a ball that has just rolled under the sofa to be regarded as 'present' to us, though we cannot actually see it? There is no sense in struggling with such questions.[1] Fuzzy boundaries are pervasive and unavoidable.

What we have next to consider is the onset of this mode in individual lives. When does it begin? What kind of form does it initially take?

Clearly we may not reasonably talk of a locus of concern before some capacity for concern is present. So a pertinent question must be: when can we say that such a capacity first normally starts to manifest itself in a human life?

It may be helpful at this point to recall the distinction between *locus* and *focus* that was drawn on p.12. Within a given locus, very many different foci of concern are possible. The immediate here and now may seem to be relatively restricted; but it is rich in possibilities for the engagement of our concern, and our minds are certainly not able to deal with them all at once. So concern implies selection. It implies at least some capacity for discrimination, some measure of focusing. If we could not distinguish one thing from another, how could we be concerned with anything? We have then to try to decide at what point in life a child first has a mind that is capable of concerning itself with things in some sort of controlled and organized way.

The difficulty of knowing what a baby can perceive, think or feel is evident enough. None of us can remember; and we cannot use language to put questions to an infant. Thus for a long time it seemed to be impossible to discover anything about very young minds at all.

The lack of firmly based knowledge did not, of course, stop people from speculating. Nor did it stop them from turning speculations into assertions and building large theoretical structures on such insecure foundations. We shall have to consider some of these more fully later, but for the moment we need only note that most theorists in the first half of the twentieth century did not attribute to very young babies enough by way of structured perception and the capacity for thought, emotion or directed action to make it reasonable to regard them as capable of concern about anything. However, there were great surprises to come when at last ways began to be found of allowing babies to demonstrate the extent of their mental powers.

A major breakthrough occurred during the 1950s when Robert Fantz devised an experimental technique which, superficially, was simple and obvious to the point of triviality: he put two cards, each with a different pattern on it, at a suitable distance in front of a child's face and he simply noted the proportion of time that the child spent looking at each of them.[2]

The great advantage of this method was that it used a response – that of looking from side to side – which babies are able to make quite soon after birth. So it became possible to discover something about how babies see the world when they are only weeks old.

Fantz found that the fixation times could differ markedly. Eight-week-old babies, for instance, would look significantly longer at a bull's-eye than at an array of stripes; and by four months they would look longer at a face-like pattern than at a bull's-eye. This kind of finding justified two conclusions at once: the babies could see the differences; and also they had preferences. They found certain patterns more attractive or *in some sense* more rewarding than others. It is hard to realize now how exciting this discovery seemed at the time.

It is, of course, a weakness of the method that these two conclusions cannot be separated. If results are positive, both conclusions are justified. If results are negative, there is no way to tell whether this is because no difference is seen or because no preference exists. If the babies could see the differences but had no preferences, this would yield the same results as if they could not see the differences at all.

However, since positive results were obtained it was established that the visual world even for a tiny baby is not just a hazy blur, a vague fog, a totally confusing jumble, as had previously been thought likely. In other words, babies perceive in a structured way. And, within the initial structured world, some things receive more attention than others.

There is more than one possible view of what this second conclusion implies. One way to think of it is to say that the child finds more visual interest in one pattern than in another and therefore *chooses* to look at it longer, just as an adult might find one painting specially satisfying and so might stand longer in front of it in an art gallery exploring it. On another interpretation, however, the babies are not so much choosing a pattern as being 'captured' by it. This second argument rests on the idea that infants are drawn automatically to certain features of the world, those that are specially effective, perhaps, in stimulating their nervous systems. For instance, Marc Bornstein found that babies tend to look longer at more saturated colours than at less saturated ones, and longer at horizontal or vertical lines than at oblique ones; and in both cases there is some evidence that the preferred stimuli (the stronger colours, the horizontal and vertical lines) produce greater activity in the visual system of the brain. So Bornstein argued that the child is not so much exploring the stimuli as being seized by them.[3]

Bornstein's claim, then, was that the child's behaviour is under very direct control from outside. However, we should notice that the baby does not by any means spend *all* of the time looking at the preferred object. Thus nothing so automatic and predictable as a reflex is involved, even if the stimulus that is looked at longest is, in a quite direct neurological sense, the most stimulating one.

Also there has now been a great deal of further research; and the conclusion which is emerging clearly is that, as Eleanor Gibson puts it, 'neonates explore events'.[4] She considers three basic questions. First, is the neonate really exploring the world – that is, a world conceived to be 'out there'? Then, if so, is the exploration genuinely directed by the infant, not just a matter of compulsory response to stimulation? And, finally, does the early activity have important consequences for later development? Gibson demonstrates, to my mind convincingly, that in the present state of knowledge the answer to all three questions seems very likely to be 'yes'.

What young babies can do is, of course, limited and not always very efficient; but they turn out to be capable of kinds of exploration that go far beyond what casual observation might suggest. Some of the most convincing evidence concerns a kind of 'finding out' that involves not just looking but doing – producing effects on the world. Surprisingly, the appearance of this ability in infancy has proved quite amenable to study, given enough ingenuity on the part of experimenters.

In ordinary circumstances babies do not have much opportunity to control the world. There is a very restricted range of action that they can take. For instance, if they cry when food does not appear, this may indicate not a failed expectancy but simply a physical discomfort. So to discover what babies are capable of in this respect we must contrive special situations in which their restricted powers of movement and manipulation no longer set limits to the activities of their minds.

Various devices have been constructed for this purpose. A useful one is a mobile, set up so that the baby, lying in her* cot, can control it by movements of her arms and legs, or even by

---

* I think that the problem of which pronoun to use in referring to 'the child' will eventually be solved by the general acceptance of 'they' as both singular and plural. There is, after all, the precedent of 'you'. But my ear still insists on taking 'they' as exclusively plural, so my solution in this book is to use 'she' and 'he' in alternate chapters.

movements of her head on a sensitive pillow. Alternatively, apparatus can be set up that makes a light display go on whenever the baby looks, say, to her left or when she makes a specified sequence of head turns. Many other means of giving a baby unusual amounts of power over what happens also exist. Even the sucking response can be used if the teat is linked to some visual or auditory transducer. Kalnins and Bruner, for example, so arranged things that babies, by their sucking, could bring a picture into focus or send it out of focus.[5]

Studies of these kinds have yielded results that are quite clear: babies in the first half-year of life can quickly learn to exercise effective control. What is more, they seem pleased when they do so. For instance, John Watson, working with a mobile, reported that 'vigorous smiling and cooing' appeared on about the third or fourth day of exposure. And Hanus Papoušek observed signs of distress when control and prediction failed – when the lights in a display no longer went on as the infant had come to expect them to do, or, in other words, when the baby could not understand it.[6]

It is important to notice that in these studies the children were not given food or toys to urge them on. There was no reward for them except the success of their own efforts. In this kind of experiment it is generally found that activity levels fall once competence has been achieved. Papoušek's infants, for instance, did not in general continue switching on the lights once the problem of discovering how to do it had been solved. He concluded that the interest was not in seeing the lights, but in the establishing of effective control.

Some of the implications of this conclusion need now to be considered. If a baby gets satisfaction from learning to predict events in the world and to control them, then presumably she has some conception of a world 'out there' to be controlled. And by the same token presumably she has some conception of herself as a controlling agent. This, however, runs counter to what was for a long time the orthodox opinion among developmental psychologists. The received view (among those willing to talk of self-knowledge at all) was that the baby does not know herself because she has not differentiated herself from the rest of the universe. This was the claim made by Piaget and also by classical psychoanalytical theory. Freud argued that neonates make no distinction between self and other and recognize no outer reality. And Margaret Mahler, following in this tradition, asserted that the first few months of life are a time of 'symbiosis' between mother and infant – a kind of continuation of prenatal life during which the infant is not, psychologically speaking, a separate person at all.[7]

Piaget was no doubt influenced by the Freudian idea; but he used it in different ways as the foundation for his own arguments about the later course of cognitive development. These two men, with their powerfully constructed theories, have been so influential that when they agree the matter is apt to appear to be settled.

Piaget's claim that the indifferentiation of self and not-self is the normal initial state rested on evidence of a kind that at first seems very strong. His research had indicated that when a baby is playing with a toy and the toy is removed and covered with a cloth, then, up to the age of eight or nine months, the baby makes no attempt to pull the cloth away and reach the toy again. Piaget took this as proof that the child has no idea that she lives in a universe of enduring things, *one of which is herself*. He argued that the failure to search for a hidden object implies that, for the child, this object has ceased to exist – that out of sight is indeed out of mind; and, further, that the development of the notion of inde-pendently enduring objects – the 'object concept' – is inextricably bound up with the knowledge of one's self. On this view, one comes to know that one exists as a separate being to the extent that one distinguishes one's self from the rest of the universe. The claim is that the two ideas necessarily develop together.[8]

Now, if this were true, then clearly the children's satisfaction in learning to move a mobile, or switch on lights, or bring a picture into focus could not derive from an enhanced sense of their own competence; for how could there be a sense of competence if there were no sense of self? Likewise, how could there be pleasure in learning to make a mark on the world if there were no know-ledge of any world out there to be dented? If we accept the

Freudian/Piagetian idea of initial lack of differentiation – or 'complete and unconscious egocentrism', as Piaget sometimes called it – then the evidence obtained by Papoušek and others has to be interpreted in different ways. For instance, Papoušek claimed that the babies in his experiment were matching incoming information about the world against an inner 'standard', which amounts to saying that they already had some kind of mental model of what the world is like. But if they had not distinguished themselves from the rest of the world that would not seem to be possible. It is therefore important to ask whether evidence for the existence of this early undifferentiated state is as strong as Piaget believed it to be.

Discussion of the object concept is held over till the next chapter, since it raises questions that bear on the onset of the line mode. However, there is another issue that is perhaps even more fundamental. Whatever the case about the development of belief in a universe of solid enduring 'things', is it really true that this is inseparable from belief in a universe distinct from ourselves – that we can't have one without the other?

Leaving Piaget's special arguments aside, there is no doubt that to an adult, on first reflection, the two ideas may seem very closely linked. We are so deeply committed in our ordinary thinking to the notion of a world of objects which remain 'the same' across time that we hardly consider the possibility of worlds differently constituted (unless, of course, we are engaging in philosophical speculation). But the two beliefs are not in principle necessarily linked. And it might well be that they are acquired separately. A baby might first come to a sense of her self as distinct from something 'other' but interacting with it, and then gradually build up a notion of how that 'other' is constituted and how it is to be dealt with.

It is hard to be sure, but I am inclined to think that some sense of self comes very early.

There are good reasons for supposing that, from early infancy, we should distinguish the perception of our own functioning bodies from the perception of the world outside. For one thing, the contrast between the feelings we get when we move the environment and those we get when it is the environment that moves and we who merely experience the movement may be quite primitive; for much simpler nervous systems than ours can detect this difference. Johnson argues that an organism unable to distinguish its action on the world from the world itself 'would be unable even to begin to adapt to reality and to organize actions'.[9]

Also there is now experimental evidence that four-month-old babies are sensitive to the difference between watching something that moves while they stay still and watching something that stays still while their own position changes. They seem to see an object as stationary when its displacement relative to them results from their own movement – just as we do. But further than this, they see an object as moving when it really is moving, even if they themselves are moving to a corresponding degree – again, just as we do.[10]

Another relevant point is this: it is a characteristic of the normal feeling of one's own body as functioning that it provides us with a sense of continuity – of derivation from the past and direction towards the future. Notice that this sense of continuity need not depend on the elaboration of an extended personal past containing specific remembered events or on any notion of an extended future. Only a sense of states continually yielding place to other states is entailed. By contrast, incoming information about the world can – and often does – yield abrupt discontinuities.

Thus the sense of a differentiated and enduring self might in principle precede any conviction about enduring things in the world beyond. After all, many percepts coming 'from outside' are not percepts of enduring things at all. There are shadows and reflections, there are smears of milk or dust that may be wiped away, vanishing effectively for ever. There are things that burn or dissolve before our eyes. So it might well take some time to discover when a disappearance from sight is likely to be followed by a reappearance and when it is not.

These, however, are not the only relevant considerations. For we now have a whole new body of evidence bearing on this topic – evidence not available to Piaget when he made his claims.

Until recently it was widely assumed that a newborn baby received sensations of touch, of sight, of sound, and so on, that were initially unrelated across modalities. Thus, for example, knowing what an object felt like would provide no information about what to expect if it were seen. On such a view, early experience was fragmented. Structured organization was not given, it was something that had to be achieved: achieved by long hard work according to a constructivist theory like Piaget's or by more passive, but still time-consuming, associative processes according to the behaviourists. These two kinds of theory, so different in some ways, shared the fundamental idea that relationships between inputs from the various sensory channels were not provided for us ready-made but had to be established step by step.

We now learn that this assumption is, to say the least, highly questionable. For some time evidence has been accumulating that early perception is pervasively 'supramodal' or 'amodal'. That is to say, it somehow transcends a specific sensory channel. For instance, in a much-quoted study Meltzoff and Borton gave three-week-old infants dummy nipples to suck, covering the babies' eyes so that they could feel the shape of the nipple but could not see it. Two distinctive shapes were used, each child having experience of one only. When the babies were later shown two nipples side by side, they looked more often at the nipple of the shape they had previously come to know through sucking. Thus response to the new visual input was influenced by prior experience in the modality of touch.[11]

Gibson comments that this research has proved hard to replicate, but she suggests the reason may be that infants who are only a month old attend rather poorly to visually presented shapes. She reports that another study of her own, using movement, tended to support the same general conclusion: early perception has an amodal quality.[12]

There is still much to be discovered about the perceptual base from which the newborn starts and how later developments proceed from there. Gibson's review gives some idea of the complexity of the issues. But it now seems unlikely that we will ever return to the old notion that early experience is a kind of bombardment by unrelated sensations.

Stern reviewed the evidence for amodality as it stood in 1985 – evidence, for instance, that babies can detect correspondences between levels of intensity of light and of sound; and that they can recognize the similarity between a rhythmic pattern which is heard and one which is seen.[13] Speculating about the quality which all this confers on an infant's experience, Stern suggested that there will arise feelings of *déjà vu*. The present will be imbued with a kind of familiarity, even in the absence of anything one could call anticipation. Another way to put this is to say that supramodal perception must help the child to feel 'at home'.

One kind of 'at home' feeling that is of special importance may arise from the perceived correspondence of sight and sound when someone looks at a baby and speaks. It turns out that, from an early age, babies are sensitive to the co-ordination of these two kinds of stimuli over time. Barbara Dodd has found that babies between ten and sixteen weeks of age know when speech sounds and lip movements are in synchrony and when they are not.[14]

Dodd compared two conditions: one in which the sights and sounds corresponded in the normal way and another in which the sounds were artificially delayed by 400 milliseconds. When these two conditions were presented in alternation (a minute of one followed by a minute of the other) it was possible to assess the amount of attention paid to each. If the children looked away, played with their clothes, fretted, or showed any other signs of distraction, observers pressed an 'inattention button' to record the duration of the distracted state. There was a high measure of agreement about this between two independent observers; and the outcome of the study was the finding that the babies spent significantly more time attending to the normal stimulus, where sight and sound matched one another, than to the discrepant one. (This was true for the babies as a group and for most of them individually, though there were quite striking differences between infants. One baby attended to both stimuli for almost the whole time.)[15]

The interest of Dodd's work goes beyond the question of whether babies can perceive the correspondence of sight and sound, for it bears also on the question of whether lack of correspondence troubles them. Are they upset if speech sounds and lip movements are out of phase? Some of Dodd's descriptions suggest that the answer is 'yes'.

Lynne Murray, working with babies aged between six and twelve weeks, looked specifically for evidence of distress when various anomalies or 'perturbations' were introduced into the normal pattern of communication between child and adult. She did not concern herself with speech sounds and lip movements but with more general facial and vocal responsiveness. For instance, in one study she arranged that brief periods when mother and baby would 'chat' normally should be interspersed with others when the mother would keep her face blank. When the babies' behaviour was studied closely many highly significant differences between the two conditions emerged.[16]

In the normal condition the baby typically looks for long stretches of time at the mother's face, makes many movements of mouth and tongue and often smiles. Frowns and grimaces, if they occur, are transient. There is a strong impression that active attempts at communication are going on.

In the 'blank-face' condition, on the other hand, Murray found that smiles diminish, frowns and grimaces increase, the baby tends to look away more often and to play with her own clothes or her own hands. She seems troubled; and communicative efforts, when they appear, look hesitant and uncertain.

On this evidence, then, a baby at six weeks of age already knows the difference between a responsive mother and an unresponsive one and does not seem to like to encounter the latter.

In further experiments Murray compared the responses of child and mother in two situations: one where the partner, though not in the same room, appeared live on a television screen, so that normal 'conversation' was possible; the other where the television picture was in fact a replay of the immediately preceding episode. The baby was thus seeing a perfectly normal range of friendly expressions on the mother's face. All that was lacking was genuine interaction.

It is highly interesting to learn that this last condition appeared to trouble the babies as much as did the blank face. Once again they tended to look away, to grimace, to frown. And it affected the mothers too, when the same trick was played on them without warning. Their speech changed in significant ways, so that they made more remarks that seemed negative in tone, made more references to themselves, and so on. This last finding implies that the infant really contributes in important ways to the success of the exchange from the mother's point of view.

Murray concludes that 'the manifestations of protest or solicitation, and of distress in the face of breakdown in communication lend weight to the view that the infant has intentions to engage in reciprocal, harmonious interactions'.

Fortunately, these intentions are often well realized. As Colwyn Trevarthen has consistently reminded us, infants, given any kind of reasonable chance, appear to experience much fun and happiness in their relations with others. Trevarthen argues that in our culture we have come to look on emotion in essentially negative ways that are misguided and detrimental.[17]

The evidence we have been reviewing shows the infant to be in many ways highly competent — sensitive, inquiring, an efficient learner. It also reveals clearly, as do many studies, that infants are specially sensitive to other human beings. That is, within the here/now locus of concern the *focus* of concern is very often on relationships with people. This fact may seem unsurprising; yet it has not always been given the prominence that it must have if we are to understand human development. We have already seen that the inanimate world also evokes much interest from children at an early stage. However, the child's experience of this world is to a large extent mediated by other people; and it is a distinctive thing about our nature that our learning is not achieved alone.

Learn we do, however, from the beginning. And the extent of our early competence may seem to raise some difficulty for the argument I have been developing. Can it really be that, for such a person as the infant turns out to be, all experience is confined to *now*? After all, does learning not entail remembering?

Consideration of the nature of infant memory will come in the next chapter. However, to anticipate, it seems likely that babies under eight months or so have as yet no memory of an extended, ordered past stretching behind them. Though past events can already strongly influence present experience, specific happenings are given no locus in remembered time. Thus event B is not thought of as having happened between A and C. Nor do babies appear to have any conception of a future lying ahead, except for the brief anticipations which are made possible by the fact that the experienced present moment is never vanishingly brief but has its own kind of extension, guaranteeing continuity. If it were not so, then there could scarcely be any responsiveness to anything. And there is a great deal of responsiveness right from the start. We must therefore postulate that, even in early infancy, there is some brief extension backwards and forwards, so that the context in which the mind functions reaches to what has just been and to what is just about to be. If we think of 'now' as a point, then it is not like the idealized extensionless point of the mathematician but rather like a moving spot of light – and fading out at the edges, not in general sharply bounded.

To get some sense of what this early point-mode experience might feel like, think of listening to a concerto which is familiar but not well known. When and where did you hear it before? You don't know. You must have heard it somewhere, though, or your present experience would be quite different. At the start of the first movement you cannot anticipate the second movement at all; but as the first movement closes the opening bars of the next one arise in your mind. Much infant experience is probably like this except that the question: 'When did I hear that before?' cannot be asked – and not just for lack of words but for lack of an appropriate conception of structured time.

The main aim, up to this stage in the argument, has been to understand the general nature of the mental base that is already established when the line mode starts to appear on the scene. But before going on to that next major event, it will be as well to consider briefly some of the ways in which the point mode manifests itself later in life, if only to stress here at the start that there is nothing intrinsically primitive or lowly about its activities. For instance, language is learned as an outgrowth of its functioning from the second year onwards. Children learn language by being spoken to about things or events around them – things that serve some immediate purpose in their lives, things that interest them, things that matter. And when they themselves first speak they choose the same kind of topic. Language, when it begins to appear and in much of its later use, is embedded in a context of current happenings directly experienced, and in the accompanying inner context of feelings and aims.

It should also be recognized that, although direct concern with what is here and now is accompanied in early infancy by narrow temporal awareness, this does not have to be so. The restriction, if it arises, is a function of the individual mind at a given stage in development. So when an adult concentrates on a skilled task, such as upholstering a chair, there is the absorption in the moment that is typical of the point mode, yet there is a great reliance on past experience and a well-formulated goal that is some way ahead: the finished chair.[18]

Like the crafts, the visual arts belong to a large extent in the point mode. Sculptors shape their materials here and now as they work. Painters are concerned at every moment with the immediacies of their pigments as they apply them to the canvas. And the direct appreciation of art is a point-mode activity. Here 'direct appreciation' means looking at the work of art and is to be distinguished from certain kinds of critical reflection.

The case of music is more complex. The composer seems to be less dependent than the visual artist on any direct interaction with what is immediately present. Beethoven could even compose great works – some of the greatest of all time – when he was deaf. However, the performers of music, and their audiences if attentive, are clearly functioning in the point mode.

As for science, although the enterprise of scientific inquiry would be impossible without the later, still to be considered, modes, we must recognize that the point mode has a crucial role to play here too.[19] For one thing, the curiosity that powers science has its source in direct observations not well understood and so giving rise to spontaneous wonderings. This is true at all ages but here are two fine examples from young children. The first concerns Jamie, who was three years eleven months old at the time the following dialogue took place. Jamie was standing with an adult in a lane beside a house in the country. A car was parked on a concrete base nearby. It was a warm, dry day.

JAMIE: Why is it [*the car*] on that metal thing?
ADULT: It's not metal, it's concrete.
JAMIE: Why is it on the concrete thing?
ADULT: Well, when it rains the ground gets soft and muddy, doesn't it?
   [*Jamie nods, bends down and scratches the dry earth.*]
ADULT: So the wheels would sink into the mud. But the concrete's hard, you see.
JAMIE [*excitedly*]: But the concrete's soft in the mix! Why is it soft in the mix?

The adult, who had not thought of this in the context of the conversation, was thrown into some confusion and was quite unable to answer Jamie's last question satisfactorily.

The second example comes from a conversation between a small girl and her mother, overheard in a bus. The child seemed to be around four years of age.

CHILD: Mum, where's the moon? Why is it not there tonight?
   [*The sky was dark, no moon was visible.*]
MOTHER: The moon's gone to its bed.
CHILD [*indignantly*]: Moons don't have beds!

This response was altogether too much for the mother. After a few seconds of silence she said helplessly: 'Well, where do they sleep then?' And she rapidly changed the subject of the conversation.

In both examples, the children are spontaneously wondering

about something directly seen. Thus both start from a present-moment locus of concern, though the attempt at understanding soon leads them to draw on what they remember or what they already know: the concrete mixer with the soft, churning stuff inside; the fact that moons and beds in no sense belong together. Both children, for these reasons, reject the explanation that the adults offer them and are presumably left in the end still wanting to know.

However, this kind of spontaneous wondering is not the only role for point-mode functioning in science. There is another much later subcategory of a very special kind. An important part of scientific inquiry is the attempt to make observations and conduct experiments in ways uninfluenced by personal goals other than the goal of finding out. That is, there is an effort towards what we call 'objectivity'. This is point-mode functioning in the sense that we are concerned with what is happening here and now: we are observing and, in experimental work, we are also taking action.

In this special sophisticated kind of perceiving and doing the original unified quality of mental life is given up in favour of a splitting off of personal feelings and aims, so that there remain only the austere aim of discovering 'the truth' and certain relevant emotions that tend to accompany this, mitigating the austerity. The separation needed is hard to achieve or maintain and demands considerable maturity and control. (See chapter 10 for a further discussion of this topic.)

Activities in which the locus of concern is the world as we meet it and deal with it here and now must clearly remain of great importance throughout life. Our survival depends on the adequacy of these encounters. However, once we achieve competence we may not need the whole mind for them any more. Often, to use contemporary jargon, familiar activities can be assigned to 'subroutines' – or we can 'fly on automatic pilot'.

Thus as we grow older – and particularly if we live in a fairly stable and prosperous society where basic needs are easy to satisfy – there may be fewer and fewer occasions when we are really absorbed in the present moment. In *The Silver Spoon*, one of the novels in his chronicle of the Forsyte family, John Galsworthy describes Fleur's son, Kit, taking a bath at the age of fourteen months in the following terms: 'His kicks and crows and splashings had the joy of a gnat's dance, or a jackdaw's gambols in the air. He gave thanks not for what he was about to receive, but for what he was receiving.'[20]

But when an adult takes a bath it is commonly a very different matter. The hands regulate the temperature, the limbs feel the warmth, but the mind is often 'far away'. No resemblance to the gnat's dance or the jackdaw's gambols remains. As adults we are normally no longer fascinated by the gushing, splashing water, or the movements of a plastic duck on the surface, or the effects of squeezing a sponge. Point-mode activity continues, as it must: we are indeed concerned not to have the water too hot or too cold for comfort, and so on. But other kinds of mental activity go on at the same time. That is, other modes limit the scope of the point mode – limit the intensity and the delight of absorption in direct experience, to the point where we may lose this altogether, or at least for long stretches of time.

We are now in a position to return to the question raised at the beginning of this chapter: at what point in life does a child first have a mind capable of concerning itself with things in some sort of controlled and organized way? And if we are still not able to answer quite definitely by saying: 'At birth', or: 'From the nth day', we can at least now confidently reply: 'Very early, certainly by the end of the first two or three months, possibly sooner.' The evidence that has been considered – and much more that has not been mentioned – clearly justifies this conclusion.

Stern suggests that there are distinctions to be drawn between different 'senses of self' that emerge as a baby grows older. He claims that the first two months of life bring the development of what he calls the 'emergent self' – a sense of self that derives from 'the *process* of emerging organization as well as the result' – that is, from the experience of relating experiences one to another. There follows, from two to around eight months, the development of the 'core self' – a sense of self that is coherent, firmly distinguished from what is other, but not yet informed by an awareness of other *minds*. These developments, then, occupy the whole of the time when the point mode is the only one available.[21]

I find Stern's arguments about this distinction persuasive. I think it leads, if we accept it, to the decision that the point mode begins as the core self is established. This, at any rate, is the properly cautious view to take at present. It avoids claiming that in the present state of the evidence we may confidently speak of a locus of concern from birth. At the same time it recognizes that the emergence of behaviour which seems undeniably expressive of concern with the world – concern to explore it, to engage with it – is not long delayed.

However, to establish the existence of a capacity for concern is not yet enough. We must also consider whether the four components of mental activity that are used in specifying the modes are present. That is, can we say that a baby of two months or thereabouts already acts, perceives, thinks and has emotions?

A positive answer concerning all of these is strongly implied by the evidence already presented. But it may be worth adding some further comment on the question of emotion, which is the most controversial. We have seen that Dodd and Murray both report observations of behaviour which seems expressive of emotion – as do very many other experimenters and observers. For instance, Louise Kaplan offers us a description of the way a baby rages: 'He kicks and shakes the bars of his crib with unmistakable fury. He bites and pinches – himself, his mother, his pillow. He turns red in the face, blue in the face. Some babies rage so furiously that they lose their breath and faint . . . It is an awesome sight.'[22]

To someone watching this kind of display the impression that the baby *feels* anger is overwhelming. Yet people have wondered whether this might be illusory. How do we know that the baby feels anything at all?

There is, of course, a sense in which we can never be certain of

what anyone else, infant or adult, feels. But adults and older
children can tell what they feel. Babies cannot. Also it has been
argued that the nervous system is not at first mature enough for
emotion. Thus Melvin Konner, speaking of the key pathways of
the limbic system, says that 'there is thus a real sense in which the
"stream of feeling" cannot be said to be properly functional until
they [the nerve fibres] are at least substantially myelinated'.[23] Konner
is trying at this point to account for the widely reported absence
of fear of strangers during the first half-year of life, but he writes
as if the neurological research he cites had more general implications. However, LeDoux claims there is now overwhelming
evidence that what he calls 'affective computations' depend
essentially on certain neurons, or neural networks, in the amygdala,
deep in the temporal lobe. These receive inputs not only by way
of the cortex but also directly from the thalamus. Thus it is
possible for there to be a very rapid process of value feeling in
certain relatively simple situations. LeDoux speculates that the
thalamo-amygdala circuits may have a very important part to play
in early emotional experience because they mature early and can
operate before the neocortex and the hippocampus are fully
functional.[24]

On this view the present state of neurological knowledge
presents no bar to acceptance of the belief that infants do experience
certain varieties of emotion in the early months of life, as observations of their behaviour most powerfully suggest. So it looks as if
the four components are present.

But now another question arises and it concerns the manner of
relationship of these components. We can intellectually discern
them as we reflect on infant powers, but are they separable in
infant experience? Can infants, for example, perceive or think in a
cool, detached, dispassionate way? Or can they think, or feel
emotional, about something they are not currently perceiving?

It seems highly likely that the answer has to be 'no', that
whatever is the immediate focus of concern is experienced by the
baby as a kind of total immersion in a way that admits no
separation of percepts, thoughts, feelings and actions. These are

intertwined – perhaps one should even say fused – so that the whole of the personal life is unified, immediate, spontaneous.

It is hard – perhaps impossible – to prove this; but I know of nothing to suggest a contrary view. And Stern has this to say:

Infant experience is more unified and global [more, that is, than academic analysis might suggest]. Infants do not attend to what domain their experience is occurring in. They take sensations, perceptions, actions, cognitions, internal states of motivation and states of consciousness and experience them directly . . . All experiences become recast as patterned constellations of all the infant's basic subjective elements combined.[25]

If the argument so far is sound, we have established that babies can operate in the point mode from at least the age of two or three months. They are capable of concern. They perceive, think, feel emotions and act, though these functions are almost certainly not so distinct and separable as the words we give them in analytic reflection seem to imply. The baby's experience is a seamless cloth, rich in strands but tightly interwoven.

4

The Onset of the Line Mode: Remembered Past and Possible Future

'The optional character of life is perceived qualitatively as vitality.'
– Susanne Langer

'. . . actual occasions are selections from the realm of possibilities . . .'
– A. N. Whitehead

It is a fact about the universe that anything which happens precludes infinitely many other events which, without it, might have happened. Thus in addition to their positive consequences events bring with them a vast, vague shadow of negative consequences: things which otherwise might have been. For example, every one of us exists at the expense of countless unborn sisters and brothers.

Actual occasions, in Whitehead's terminology, are possibilities that have achieved ingress into actuality. These are the realities that we encounter and with which we must deal all our lives long. We ourselves in the moments of our lives are among their number.

However, our minds are apt also for consideration of the merely possible: what might have been, what yet might be.

Constraints on the possible exist because of the relatedness of things. The occurrence of A permits – maybe even necessitates – the occurrence of B. At the same time, A precludes C. Thus compatibility and incompatibility arise; also impossibility and necessity.

But how does the understanding of these things arise in our minds? The world as presented to our senses does not invite the apprehension of them. Occasions which have achieved ingress just are as they are. They do not directly reveal their own status as realized possibilities; and they do not display the other possibilities to which they stand in relation.

The best time for becoming aware of possibility is before the ingress into actuality has taken place. To such a time we have one direct and primitive means of access: through our awareness of our own impulses to action. Whenever we consciously entertain two or more impulses and wonder: 'Shall I do this or that?' we have a sight into the realm of possibility. And whenever the impulses are such that the realization of one would preclude the realization of the other we encounter the hard truth that impossibility is a feature of the universe. We are forced to choose.

To the extent that an animal's actions are determined in advance by heredity, such conflicts of impulse will not occur. And to the extent that consciousness and representational resources are limited, such conflicts, even if they occur, will not be experienced. But in the higher animals, where more and more depends on the initiative of the individual, the clash of irreconcilable urges will become more common. In human beings, where the possibilities for variety are very great, conflict of impulse will be frequent and inescapable. Certain impulses will have to be denied expression. And this fact will, at least some of the time, reach awareness.

The discovery that the doing of one thing may rule out the doing of another comes early in a human life. Children soon discover that you cannot in the same moment hug your teddy bear and throw it on the floor. In such a case there is no escape from the fact that one or other impulse has to be abandoned. Thus the world opposes itself to the child by virtue of constraints inherent in the very nature of space and time.

Jerome Bruner reports a series of observations of the behaviour of children aged between four months and twelve months when they were offered two or more toys in succession.[1] The emphasis in this study was on the development of motor skills, but the work also provides good evidence about the handling of conflicting impulses.

At four months there was no sign that the children could deal in any systematic way with more than one impulse at a time. When the second toy was presented they either ignored it completely or transferred total attention to it, dropping the first one as if inadvertently, without noticing what had happened. This typifies the distractibility of early point-mode behaviour: the new salient event displaces the old.

There followed a period when conflict was evident. A child who had put an object in his lap in order to have a hand free to grasp another one would show signs of an impulse to pick up the 'stored' toy again – an impulse incompatible with the intention of holding the new one. However, by the age of twelve months a smooth way of dealing with this problem had commonly been found. Indeed, the children had the whole matter so well under control by then that they would often anticipate the arrival of the next toy by storing its predecessor quickly and holding out a hand before the new one appeared. In this way they showed that a decision had been made to inhibit one impulse and express the other. They appeared to have realized already that they could have one or other toy in the hand but not both – that it was necessary to choose.

A sense of options, whether compatible or incompatible with one another, implies some ability to contemplate that which is not yet – that which only *may* be. Since, by definition, that which is not yet presents no stimulus to the senses, it can be contemplated only by a mind capable of calling it up.

This is not to say that the calling up has to be independent of stimulation. It is, after all, the presentation of the toys which induces the child to think about the options open. But the consideration of these options entails awareness of the likely *outcomes* of possible acts, and these are available for scrutiny only by virtue of the child's ability to bring them to mind.

What is brought to mind must be based, to some extent at least, on memory. We may assume that the child, in foreseeing the future, is using past experience. A familiar distinction between two kinds of remembering is now relevant. There is, on the one hand, the ability to *recognize* and on the other the ability to *recall*.

The difference between these is that, in the case of recognition, there is a present stimulus to 'trigger' the memory: we see a person, for instance, and we know that this is someone we have seen before. By contrast, recall occurs when that which is remembered is not present to our senses, when we 'call up' a memory of someone who is no longer there. It is at once evident that this must entail a greater measure of autonomy and control.

Notice, however, that recognition and recall are not so sharply distinct as they might seem. In the extreme case of true recall there is a memory not directly evoked by *any* present stimulus (but led up to, perhaps, by a 'train of thought'). In the extreme case of recognition there is the full presence of the thing remembered. In between, various degrees of 'help' may be given by present stimulation. For example, suppose that babies learn to operate a mobile by jerking a string attached to their toes; and suppose that later they are placed in the same situation but with no connecting string. They will then kick as if the string were there. This is memory certainly, but is it recognition or recall? Generally it is known as 'cued recall', but it is very close to recognition. However, the more partial or incomplete the re-presented stimulus, the closer does the memory come to recall. The most famous example of a minimal yet very powerful cue comes from Proust's great novel, *A la recherche du temps perdu*, where the taste of a madeleine dipped in tea evokes a great involuntary rush of memories from long ago.[2]

Wholly uncued recall seems unlikely to occur in early infancy, though we must acknowledge that, if it did, it would be extremely hard to detect. What is certain is that, by definition, it does not occur so long as all experience is confined to the present moment – that is, so long as the point mode is the only one available.

As for recognition, it is now well established that babies have a considerable capacity for it within the first few months of life. If a baby is shown two objects, one of them seen before and one of them new, then there is a strong tendency, regularly noted, for the child to spend more time looking at the new one. Given that the two objects are widely different and that the child is allowed enough time to look at them, this kind of experiment shows that recognition memory is already there even by one month of age. By five months, infants have been shown to recognize photographs after an interval of fourteen days, and there is much evidence for the early recognition of faces.[3]

We may even go further: it seems that newborn babies can distinguish sounds heard in the womb from others not heard before. They are reported to show a preference for their own mother's voice; and, still more remarkably, they seem to recognize the exact words of a story read to them *in utero*, for there is evidence that they prefer that version to a version with a few words changed.[4]

However, we must be cautious about interpretation. These experiments establish that past experience is affecting present behaviour. They do not establish that what is in the child's mind is precisely what an adult means by 'recognition'. We cannot be sure that the baby is *aware* of looking at a familiar object or hearing a familiar voice; and we certainly cannot conclude that there exists what Lockhart calls the attribution of pastness.[5] As we have seen, it is very likely that the baby, in the first half-year of life at least, while having a sort of 'rolling' sense of movement from immediate past to immediate future, has no sense of an extended past in which specific events can be located – and likewise no sense of a future filled with events yet to come.

What evidence can tell us when the past and the future start to figure in a child's conscious mind? The issues are complex and difficult, as always with questions of pre-verbal awareness. However, there is reason to think that memory changes dramatically about three-quarters of the way through the first year. The period from eight to ten months appears to be critical.

It is around this time that babies start to show fear of strangers. Around the same time they begin to show distress if a known person to whom they are attached goes away. Kagan and his colleagues argue convincingly that the distress depends on memory of the departure.[6] They imply that this is genuine memory for a past event: the child recalls the previous presence and the going away. The younger baby, lacking this ability to call up the past, is not troubled.

There are, of course, differences from child to child in the precise nature of the new response that occurs when the familiar person leaves or when a stranger appears. Louise Kaplan describes it as ranging 'from wonderment to panic'.[7] But some component in it of distress or wariness is extremely common. And the phenomenon is so widespread, so regularly occurring around the same age even in widely differing cultures, that specific frightening experiences cannot be invoked to explain it. The child must be afraid not of what has happened, since the most gently treated of children may respond in this way, but of what might happen. The fear is fear of an unknown future. And it suggests that a new kind of question has arisen in the child's mind – a question which, if speech were available, might be verbally expressed as: 'What is going to happen?' And perhaps also: 'How am I going to cope with it, whatever it is?'

Although it is fear of people that has received by far the most attention, research by Schaffer and his colleagues shows that unfamiliar objects also evoke a kind of wariness; and this happens at the very age when stranger fear begins, that is, from eight months on.[8] Before this age the babies studied reached out for objects regardless of whether these were familiar or not. But within the space of the next month a sharp change occurred: immediate indiscriminate reaching was replaced by a period of hesitation in which the child sat motionless, as if trying to decide what might be the consequences of action. If this is the right interpretation, then the locus of concern has moved decisively from the present into the future, and functioning in the line mode has begun. But is this so? There is further evidence to consider.

The onset of new kinds of wariness or fear is not the only sign that important changes relevant to the start of the line mode are taking place around the eighth or ninth month of a child's life. Evidence also comes from a very different source: from experiments on how babies behave when objects they are looking at suddenly disappear from view, for instance by having a cover placed over them.

The literature on this subject is massive, the findings are complex, and the arguments about how to interpret them are more complex still. For our present purpose, however, much of the debate is not relevant. We are concerned at this point only with the light that is thrown on the development of memory. In particular, what kind of memory must we suppose to be present if we are to account for the children's *successes*? It is important to make, and to keep in mind, a clear distinction between this question and the question of whether the children's failures can be attributed to memory deficits *alone*. Certain facts are well established.

During the first half-year of life, or thereabouts, children do not normally try to retrieve a toy which they have just seen being placed out of their sight – under a cloth, for instance, or under an upturned cup on a table in front of them. They act *as if* the hidden object no longer existed.

Then comes change: a baby will now typically try to get the toy back and will succeed, given that there is no doubt about where it was hidden. Suppose, however, that two upturned cups are set upon the table-top, so that there is no longer only a single hiding-place. Suppose further that a baby has seen the toy placed under cup A, and has got it back successfully. What will happen if the toy is next hidden under B with the child watching all the while what is done?

Adults tend to start by assuming that this last condition will present no special difficulty. After all, if the baby can retrieve the toy from one hiding-place, why should another equally visible and obvious one be any harder? Yet the fact is that a second hiding-place *is* harder; and, faced with this problem, children aged between eight and ten months typically make errors. They search, but often not in the right place. Having found the toy at point A on a previous trial, they tend to search at A again, even though they have just watched the toy being hidden at point B.

This is what reliably happens; but what does it tell us about the growth of memory? In the first place we may conclude that, for attempts at retrieval to occur at all, the baby must have some ability to recall that there *was* a toy which then vanished. Without memory of a recent past containing that specific event, there would be no searching.

So far, I think, the argument is not controversial. If the baby searches, the baby remembers. However, we may not go on to conclude that when the baby did not search, the baby did not remember. Something other than a memory deficit may have served to inhibit search when the child was younger. Thus the evidence so far considered does not prove that there is an important *change* with respect to memory around eight months of age. It tells us something about the state of affairs from that age onwards; but, taken by itself, it tells us nothing of what went before.

Fortunately, of course, we do not have to take pieces of evidence by themselves. The regular practice in science – as in other human activities – is to put together evidence from different sources in search of the best way to make sense of the whole.

Consider now what is to be made of the errors that arise when, having become able to search, a baby is presented with a problem involving more than one possible hiding-place. Much has been made of the 'perseverative' nature of these mistakes – of the fact that the baby goes back to the place where the toy was previously found. But notice that when only two possible hiding-places are available – as has been the case in most of the research on this topic – errors are necessarily perseverative: the child either chooses B, which is correct, or goes back to A. It appears that, when more numerous possibilities are on offer, search patterns are not always perseverative. Errors may consist in choosing a place close to the correct one rather than in returning to the place where the object was last found.[9]

The most reasonable interpretation of these findings would seem to be that the children remember well that there was an object which disappeared but they do not recall very clearly just *where* the disappearance took place. Their memory is imprecise, tenuous. They are easily confused. This interpretation receives support from evidence that it matters how long a delay is interposed between the disappearance and the start of the searching. For instance, in one study babies of nine months searched at location B successfully after a three-second delay, having previously found the object at A three times. However, a delay of seven seconds caused all the children (there were eight in the group) to look again at A rather than at B. On the other hand, by the age of ten months they could all handle a seven-second interval without error.

This sort of finding suggests a system in rapid growth. Schacter and Moscovitch believe that this is indeed the case. They argue for the existence of two memory systems, which they term simply *early* and *late*. The former, they say, brings changes in behaviour as a result of experience but does not entail conscious memory (either in recall or recognition) for specific past events. The latter does. When it is in operation, we consciously remember particular episodes in our lives.[10]

The evidence reviewed by Schacter and Moscovitch appears to imply that the early system functions virtually from birth, while the late one begins to operate in the eight- to ten-month period and is well established by the end of the first year of life. Schacter and Moscovitch think there is nothing to suggest that further distinct systems appear beyond that time. They believe that later development 'consists of integrating this machinery with other cognitive functions'. Their argument is strengthened by discussion of memory in normal adults, where both systems can be shown to be functioning.[11] It is also supported by evidence from the study of certain adult amnesics who, in searching for hidden objects, make errors very like those observed in eight- to ten-month-old children. The conclusion drawn is that, in the amnesic subjects, it is the late system which has broken down.

The functioning of the late system, as defined by Schacter and Moscovitch, is evidently a precondition for the development of the second mode. To continue the metaphor used for naming the point mode, we may think of the experiential point now extending to become a line.

In its earliest form the line mode consists in having thoughts and emotions and plans and purposes about one's own life. It looks forward as well as back; and, even when looking backward, it entails much more than conscious recollection. However, without conscious recollection it would be impossible. Consequently, as far as we can at present judge, its onset is unlikely to come before the period of rapid behavioural change that regularly occurs in the last quarter of the first year. But is that still too early? Are there more developments to come which, even if they do not entail the establishing of a new memory system, are also necessary before we can reasonably talk of the personal life as having become a locus of concern?

Katherine Nelson would think so. In the volume containing the Schacter and Moscovitch paper that we have been considering she presents data collected from a little girl called Emily during a period of four months that began when Emily was twenty-one months old. As Emily lay in bed before going to sleep she regularly talked to herself, and what she said was recorded by means of a cassette-recorder placed under her pillow.[12]

Emily was a bright child, verbally precocious, who had a lot to say. Some of her speech had to do with the stuffed animals in her crib and entailed the customary kinds of pretending. But some of it, spoken in a distinct tone of voice (a kind of high-pitched recitative register, Nelson calls it) had a different kind of theme. Emily was thinking about things that had happened to her.

Two brief extracts give something of the flavour:

> 21:13   My daddy bring my grocery . . .
> My daddy come home . . .
> So my daddy Emmy daddy don't go work.
> Emmy daddy ca[m]e.
> I don't why my daddy ca[m]e.
> So Emmy move down to outside.
> I ca[m]e see cousin and see trucks.
> I ca[m]e see rocks, dirt.
> So my daddy do washing.

[N.B.: 'Came', 'can' and 'can't' were not easy to distinguish in her speech.]

21:31   Make my bed. Probably when I wake up and probably my
         sleeping liking this bed. And Emmy felt the bed so good. Emmy
         saying Emmy felt the good bed. Emmy didn't like it. [?].
         Emmy, when Emmy fell down the bed and bring [?] and
         sometimes and make new bed for Emmy. And that be now [?]
         up. That [?] bed me. And new bed for me. And new bed for
         baby. Make new. And after make new bed for baby. I [?] and
         make new bed for the for the my little baby. Maybe. I like that
         book. Mommy that book. I don't know.

    Nelson comments on the lack of coherent structure in the first
extract which, as she points out, does not refer to a single episode
or even have a common theme. But I wonder how often the
thoughts of adults, as they drift towards sleep, have a common
theme – unless there is a focus caused by something that is
troubling them. The second extract was recorded on the night
when Emmy's younger sibling – the new baby – was brought
home. Emmy had been moved to a new room many weeks
earlier so that the baby could be near the parents, but the move
had been such a salient event that it had continued to figure
in the monologues ever since. Nelson says that the extract
quoted is a fusion of the actual events in Emmy's life with a
story that had been read to her, also weeks earlier, about a father
making a new bed for a new baby. Nelson treats this kind
of 'fusion and confusion' in the thinking of a precocious child
approaching age two as evidence for a claim that, below the
age of three or so, memory is not systematic and is only gradually
becoming differentiated from a 'general representational system'
that is more appropriately to be called 'knowledge' than
'memory' since it makes no distinction between general and
specific, distant and recent, routine and novel. This is evidently
very similar to the distinction between 'early' and 'late' memory
discussed previously.
    However, the second extract from Emmy's monologues, though
it may incorporate memories of a story, is certainly organized
around a coherent personal theme. The impression of a child
grappling with her feelings about what has happened comes

through strongly, even movingly: 'Emmy saying Emmy felt the good bed. Emmy didn't like it.' One must beware of reading too much into this. But could she be reflecting that she had pretended to like the new bed when really she did not like it at all?

We can only speculate. The evidence is too slender for more. In the present state of knowledge we are not justified in concluding more than that the origins of the line mode lie in the changes that occur around eight to ten months, but that some further time must no doubt then elapse before a clear sense is developed of a past in which specific events are firmly located and of a future in which specific events can be expected with varying degrees of like-lihood.[13]

Whatever Emmy was feeling as she spoke of her new bed and of the baby who now had her old bed, there can be no doubt that the line mode is a source of much pain. Reflection on one's life can be happy as well as sad, but it is an unfortunate fact that many of us tend to dwell more on regrets than on satisfactions, more on fears than on joyful hopes. Countless lives are rendered thoroughly wretched in this way. This is the mode of 'if only . . .' and of 'what if . . .' – of resentment, guilt and fear.

Many of humanity's most severe problems – both individual and collective – stem from the prevalence and power of the line mode and from the fact that it comes upon the scene so early, while the mind is still quite immature.

As to the interweaving of the components of mental life, this is still very close as regards thought and emotion. However, percep-tion and direct motor action, to the extent that they are concerned with what is here and now, are absent from line-mode functioning. Also the planned action, when there is any, may be positive and realistic, or it may merely be fantasy. For instance, if some past event is important to us, we may think of it and feel moved to plan a commemoration, which then may actually take place – an anniversary party, a laying of wreaths. Or we may plan to change our life-style – and do it. But frequently we indulge in plans that we have no real intention of carrying out. We may think, for instance, what we would dearly like to say to someone and yet know that the words will never be spoken. Or we may plan totally unrealistic revenge.

As we grow older, a widening of the line mode normally occurs, and this makes possible behaviour of the kind we call ethical. We start to focus concern on what has happened or what may happen in lives other than our own. These may be the lives of those we love but they may also be the lives of those we hate. In the latter case unethical behaviour is liable to follow, for this comes in the wake of hatred, though it may also come in the wake of complete indifference.

The scope of the line mode may broaden still further beyond the lives of people known to us directly. We may then be concerned about groups specified by social class, ethnic membership, religious affiliation or nationhood. A display of concern for the well-being of a group to which we belong in this way may cause us to be called 'loyal' or 'patriotic'. The same terms may also be applied to us if we show ill-will to the members of another group, though this may equally earn us the label 'racist' or 'prejudiced'. A few people manage to transcend group loyalties and show concern for the whole human race or even for all living things. To the extent that people, in all these cases, are concerned about what has happened or about what may happen, they are still functioning in the line mode.

This kind of extension of the mode is evidently a development that comes only in maturity. However, to return to childhood, it is a fact of great significance that the line mode begins so early in life. Strong emotions seem to be there from the beginning; purposes begin to be entertained not long after; thoughts with a considerable capacity for ranging over the personal life are certainly present before the first few years have passed. Thus both the pleasure and the pain that go with functioning in the line mode are an inescapable part of the experience of being a child.

As we saw earlier, all children soon discover that the world of space and time opposes itself to our desires by virtue of its inherent incompatibilities: we cannot eat the cake and still have it. This is a bitter truth, much resented. But there is another way, too, in which impulses are bound to be thwarted, and it is perhaps even harder to bear. A child quickly discovers the constraints that arise from the existence of other agents with purposes of their own.

The sort of vigour and passion with which healthy children pursue their aims is certainly the normal human state, but it produces formidable problems, given that it is essential for us to live together, with close emotional ties. For our aims cannot always be compatible with those of other people. The child asserts himself in some way that is for the mother inconvenient – that conflicts with her own purposes. She gets angry, perhaps she spanks him. The child is then unavoidably angry too, and the immediate human impulse is to hit back. There is a very strong urge to reply in kind. But frequently this impulse cannot be realized, either because the child is afraid to express it – afraid of punishment, maybe, or just of 'being bad' – or because he is simply lacking in power. An adult, after all, can forcibly pick him up and dump him in a playpen or shut him in another room, and he has no redress because of his weakness and his dependence.

Some ways of living together in society make it easier than others to be a child. And certainly some parents make it easier than others. But it is never wholly easy. Always there is fear and frustration in some degree. And the nature of line-mode functioning is such that the inevitable distressing experiences can return again and again to the mind.

Line-mode experience is, of course, essential to us. The gain from being able to survey the course of life backward and forward is immense. The problem arises from the link with emotion – emotion that may be joyful but may also be overwhelmingly painful. How are we to cope if the present moment is repeatedly invaded by painful feelings flooding in from the remembered past or the envisaged future? How can we protect the present moment from being quite overcome?

In chapter 2, starting with fear as an example, we looked at some ways of dealing with this threat to the present. And we noted then that the threat begins when we are quite young.

However, it evidently cannot arise until the line mode appears on the scene.

In spite of these aspects of early experience, it would be quite wrong to give the impression that childhood is full of gloom. There is a redeeming side to our nature: we have a rich capacity for fun.

Except in most unfavourable circumstances, this capacity shows itself quickly and consistently. Trevarthen has observed in great detail the striking playfulness of babies during the first year of life – even during the first six months – and he has charted the progress of their enjoyment of teasing, their sense of the comic and their delight in shared jokes and games. He stresses that playfulness is a critically important human means of communicating and of developing co-operative awareness.[14]

Judy Dunn has followed the development of humour in somewhat older children and her findings are similar.[15] She notes that from about fourteen months of age discrepant events and forbidden actions become great sources of amusement as, of course, do forbidden words. I recall a little girl of around three who learned to say 'piss off' and, knowing very well that these were 'naughty words', used them to great effect when the vicar came to call.

Children do not merely see the fun in things. They generate it for themselves with evident pleasure. Dunn tells of a little girl who put a potty on her head and laughed at the action. But her laughter and expression clearly invited others to share in the entertainment. The fun was social.

In general, once social situations, roles and rules are understood there is much amusement to be had from mocking them by distortion and exaggeration. In such ways children stand up to life and make light of it.

5

'Pretend Play' and Conceptual Choice

Many who accept readily enough that human beings can manipulate their own consciousness may still find it hard to believe that the skill comes very early – that children scarcely beyond infancy have the ability to modify their own ways of conceiving of reality. Yet there is little doubt that they do. Developments normal to the second year of life bring precisely the kind of conceptual autonomy that seems to be called for.

The manifestations of this new mental skill may vary greatly in their details but they have in common one striking feature: children become able to treat the world not as it is, nor even as it might become, but *as if* it were other. They engage spontaneously in what we call 'make-believe' or 'pretend play'. This ordinarily begins during the second year of life.

The prototypical form of early pretend play is that in which physically present objects are made to stand for – or serve as – others that they in some measure (but perhaps quite remotely) resemble. Two other forms that appear quite early are the attribution to objects of properties which they do not in fact possess and the use in play of wholly imaginary 'things' when in reality there is only empty space.

Alan Leslie offers a theoretical analysis of pretence in which two postulated 'mechanisms' are central. One is what he calls a 'decoupler' or 'expression raiser'. When this is active so that decoupling occurs, the child's primary representation of the world (the recognition, say, that an object is a banana) is 'raised', in Leslie's terminology, to a new mental context in which it is detached from direct perception, acquires mental quotation marks around it, and thus becomes a 'thing' in its own right – a conceptual kind of a thing. So 'this is a banana', when decoupled, is no longer a representation *through which* one sees, and knows, the real banana. It is no longer transparent. Instead it has become something *at which* one looks, something opaque. (The same metaphor of transparency and opacity has often been used in talking about what occurs in the course of language development when children become aware of their own speech and, instead of just using words, start to think about them.)[1]

The notion that when human beings look at things they regularly have an urge to act upon them will by now be familiar. Given that we seldom look and leave alone, it comes as no surprise to find that Leslie's second mechanism is the 'manipulator'. Decoupled representations are the raw material on which the manipulator gets to work. They are available for this work precisely by virtue of the decoupling. The manipulator can take the decoupled expression 'this is a banana' and change it into, say, 'this banana is a telephone'. The new, manipulated expression is then marked as *pretended*. It exists alongside – or perhaps one should rather say 'above' – the primary representation, not replacing it and not contradicting it either because the two, although Leslie insists they are 'in the same code', are recognized as different kinds of thing. Thus the status of the primary expression is in no way threatened by the pretence, which, of course, corresponds to what we achieve when we know quite well that we are pretending.

The third of Leslie's proposed mechanisms is the 'interpreter'. One of its main functions is to relate the decoupled expression to what is currently perceived to be happening or to 'anchor it to the current perceptual representation', in Leslie's terms.

This brief account does not do justice to the richness of Leslie's theorizing nor to the implications that he draws from it in regard to other aspects of development, normal and abnormal. We shall return to some of these later. At present it is relevant to notice one further feature: Leslie does not discuss any forerunners of the complex abilities that he postulates. Indeed, he expressly contrasts his aim of describing the 'underlying mechanisms' with what he calls the complementary aim of fitting the ability to pretend into a pattern of developmental change. In the context of Leslie's argument pretence just 'emerges' (and he goes so far as to say that, once having emerged, it does not develop any further). For the purposes of the present argument, however, concerned as it is with the relations between different modes of functioning, the question of forerunners cannot be laid aside; and it is worth looking back now to Piaget's earlier, and very well-known, views on the subject.

In his book *Play, Dreams and Imitation in Childhood*, Piaget provides many examples of make-believe.[2] Of these, perhaps the best known is the episode with his daughter Jacqueline at the age of fifteen months:

She saw a cloth whose fringed edges vaguely recalled those of her pillow; she seized it, held a fold of it in her right hand, sucked the thumb of the same hand and lay down on her side, laughing hard. She kept her eyes open, but blinked from time to time, as if she were alluding to closed eyes. Finally, laughing more and more, she cried 'né né' ('no no').

Notice that laughter accompanies the make-believe; and this — or smiling, at least — is typically reported. Why should it be so? It is hard to be certain. However, given the observed consistency, we may reasonably take it that the smiles or laughter have *some* significance.

Superficially, of course, the behaviour is maladaptive — or at least unadapted to the world. But this view of the matter turns out to be quite wrong.

Piaget's account of what happens centres around the concept of assimilation. Assimilation is the conservative tendency in our behaviour. It 'tends to subordinate the environment to the organism as it is' and is contrasted, in Piagetian theory, with accommodation, which is the source of change and bends the organism to the successive constraints of the environment. Assimilation and accommodation, though contrasted, are held to be closely interconnected — at first largely undifferentiated, later largely complementary.

One characteristic of the assimilative function is said to be that it leads to the exercising of skills, or behaviour patterns, once they have been acquired. Whenever this kind of exercising takes place without any aim at solving some new real problem (that is, in the absence of an attempt to respect environmental constraints) the activity lies well over towards the assimilative extreme and Piaget would call it *playful*. He believes also that this 'pure assimilation' is inherently pleasurable, and he borrows from Karl Bühler the notion of '*Funktionslust*' or 'joy in functioning'. The contrast is between 'an effort to learn' and 'a happy display of known actions'.

This kind of pure assimilative action appears well before the onset of the second year of life, and Piaget gives an example of Jacqueline's behaviour at the age of nine months which is an interesting forerunner of the example already quoted. Jacqueline was in her cot:

As she was holding the pillow, she noticed the fringe, which she began to suck. This action, which reminded her of what she did every day before going to sleep, caused her to lie down on her side, in the position for sleep, holding a corner of the fringe and sucking her thumb. This, however, did not last for half a minute and Jacqueline resumed her earlier activity.

This episode occurred during a sequence of behaviour which Piaget again considered to contain no 'effort at adaptation', hence to be essentially play. He notes how Jacqueline goes through the ritual of getting ready for sleep 'merely because this scheme is evoked by the circumstances'. But this, though it is play, is not yet make-believe. Jacqueline is not pretending. What is still lacking?

The central difference is that, in the episode noted at nine months, Jacqueline is using the real pillow. At fifteen months, in the episode quoted earlier, she uses a cloth. In Piaget's words the schema (the pattern of getting ready for sleep) now assimilates to itself (takes over, or incorporates) the new object; and this object is inadequate from the point of view of effective adaptation. Another way to put it would be that at nine months Jacqueline treats the pillow precisely as what it is. At fifteen months she treats the piece of cloth as what it is not.

To say this, however, is in no way to explain why children should begin the apparently pointless activity of treating things as what they are not.

Piaget regards this development simply as the outcome of 'the previous work of assimilation'. The symbol proper can be found 'in germ', he argues, in the much earlier tendency for patterns of behaviour acquired in one context to appear in others or to be generalized. For instance, when a baby sucks his thumb instead of the breast, this is generalization of the sucking schema; 'it would suffice that the thumb served to evoke the breast for there to be a symbol.'

So it would. Piaget goes on: 'If this evocation one day takes place, it merely continues the assimilation of the thumb to the schema of sucking, by making the thumb the "signifier" and the breast the "signified".'

One may assent to this emphasis on continuity and yet quarrel with the use of the word 'merely'. For this quite glosses over the extent of the change. Something new and very important has come upon the scene.

'Generalization' is a primitive and widespread feature of animal behaviour. Another way of putting this is to say that when an animal – even a very simple organism – has learned to make a response to one stimulus, the behaviour tends to occur subsequently in response to others that are like the first one in some way. It generalizes or spreads. Thus we may say that the perception of likeness is pervasive and fundamental. And, indeed, the adaptive value of this is clearly immense. What is entailed is a sort of primitive categorizing – a very useful sort of recognition of having encountered 'another of the same kind' before.

William James once made a famous remark to the effect that a polyp, if ever it should say: 'Hello! Thingumabob again!' would thereby be a conceptual thinker. But this won't quite do. Conceptual thinking entails a good deal more. At the very least it entails the recognition of points of likeness and *at the same time of*

points of unlikeness – the simultaneous grasp of the ways in which
things resemble one another and of the ways in which they differ.

Now if Jacqueline had only responded to the likeness between
the pillow and the bit of cloth, this would indeed be merely
generalization: the 'merely' would be justified. She might in this
case have sucked the fringe and tried to use the cloth as a pillow;
then, discovering in practice the unlikeness, she would presumably
have given up. And she would have had no cause to laugh.

Thus the unlikeness is what makes the new object inadequate
from the point of view of effective adaptation. But at the same
time this very unlikeness is precisely what makes it a possible
symbol. If the cloth *were* another pillow, it could not symbolize a
pillow. Making it into a symbol, Jacqueline is using both the
likeness and the difference. *She is using them both at the same time*.
This is a very different thing from mere generalization.

And why does she laugh? I believe – though I cannot prove –
that this laughter indicates more than just pleasure in exercising an
already mastered skill – more than just 'a happy display of known
actions' (that is, the 'getting-ready-to-sleep' ritual). It has, I suspect,
a much more exciting source. What is entailed is a vast increase in
mental freedom – no less a thing than the discovery that she is able
to choose. She can treat the cloth as a cloth – as what she knows it
actually to be, since she is aware of the differences which mark it
as 'not-a-pillow'. But she can treat it as a pillow in spite of these
differences if she chooses to do so for her own ends.

The choice, then, is not just a choice of what to do. It is a choice
of what to think and of how to conceptualize the world. Whether
or not that is the reason for the laughter, Jacqueline has in effect
made the great discovery that she can regard the cloth as a pillow,
even if it is not. Thus she can in some measure begin to take
control of her own thoughts.[3]

The account just given differs from Leslie's in various respects,
one being the difference in aims already noted, another being his
explanation for the smiling, which he takes to have the mainly
social function of inviting someone else to share in the pretending.
It seems likely that such an invitation is part of what goes on,

especially if we take account of evidence that children can engage in one form of interpersonal pretending – namely teasing – from the age of about nine months and that teasing is always accompanied by smiles or laughter. However, it is also well established that babies smile at even younger ages when they master some new skill.[4] These two reasons for smiling are not mutually exclusive when we reach the kind of achievement Leslie is considering. But however the smiles are to be explained, it is integral to Leslie's theory that the forms of pretence which develop during the second year of life entail the ability to treat thoughts as objects to be manipulated, which is entirely consonant with the conclusion we have just reached in a somewhat different way.

The new skill is an essential prerequisite for the functioning of the third mode, which is called the construct mode and will be discussed in the next chapter. However, the ability to make-believe seems initially to serve other purposes.

Some of these purposes are no doubt concerned with the extension of control. Just as the young baby sets about learning how to manipulate the physical world, so now the older child turns to the business of handling the mind.

It then seems likely that once a measure of control is established and there is some choice about how to conceive of the world this power is available for a range of uses. So why not use it as a defence against unpleasant experience – as a way of making reality more bearable?

However, if the skills of make-believe are to be used effectively to this end one thing has to change. As Leslie correctly says, when a child pretends that a banana is a telephone she knows quite well that it is nothing of the kind. The primary representation keeps its status. The difference between what is real and what is pretended is maintained.

This difference is exactly what must be sacrificed if the consciousness-changing defence mechanisms are to work. The new, preferred way of representing reality must take the place of the old, more nearly veridical one, at least to the extent that only the new representation is acknown. (Freud would of course say that the old one stays on in the unconscious mind.) When this happens, then in Leslie's terms 'representational abuse' has occurred, which is to say that, following the decoupling and the manipulation, there has been a further act of substitution: one representation has effectively ousted the other, so that the two no longer comfortably coexist. Only the last process would then distinguish the defence from the make-believe play; but the distinction is evidently a crucial one.

The conclusion to which the argument leads is that, before children are two years old, they have the *capacity* to start establishing defences of kinds that entail changing their own consciousness. It would be hard to exaggerate the significance of this fact for an understanding of the human condition.

Of course, it is one thing to have a capacity and another to use it. But, given the emotional pains of childhood, it would be odd if powers apt for defence against these were to go unused for long.

It is generally not easy to observe the process of defensive 'text editing' as it occurs. Evidence of it is usually retrospective, entailing inference about what the editorial activities must have been. But here is a striking instance, observed at the time of happening.

When Julie was two years four months old, her cousin Rob had a distressing accident: a paving slab fell on his foot, breaking three of his toes. Julie at the time had a younger brother, Philip, only one month old, of whom she was very jealous. There had been incidents when she had tried to hit and scratch him.

The jealousy was normal, for Julie was an only child until Philip arrived, the sole recipient of her parents' attention. Philip's appearance on the scene was a shock and a source of much pain. Reality was suddenly very hard to bear. A short time after Rob's accident, Julie was heard telling someone that a stone had fallen on Philip's foot and that his toes were very sore. She was not simply mixing up the names. When challenged, she insisted that the accident had happened to her brother.[5]

What was Julie achieving by changing the record in this way? Was she getting rid of the unbearable idea that the small intruder was thriving? Was she finding a safe way of expressing her own impulse to hurt him? Both, very likely, at the same time.

Freud's central theme, in speaking of defence, is that the aim is control of dangerous instinctual impulses. He does, however, also speak of coping with unbearable ideas, as well as with the perceptual content of experience, and so on. And, of course, when a conception has been effectively modified, related impulses are apt to change too.

Most of the discussion so far has been concerned with modification of ways of thinking: with what is acknown. But something more direct needs now to be said about the problem of impulses that, for one reason or another, are denied expression.

We have seen that impulses cannot all be realized in action, both for logical reasons and because of the constraints of living with other people. But when an impulse is denied expression what happens? Does the impulse simply cease to exist, die away?

As far as casual observation reveals, that is indeed how it may seem. And some blocked impulses no doubt do fade, so that it is as if they had never been. But others, denied immediate discharge, find expression by indirect means.

There seem to be three main ways in which blocked impulses to action can achieve some kind of alternative realization. They may affect the body of the agent, often doing harm; they may affect the dreams and fantasies of the agent; or they may issue in overt behaviour after all, but in a different setting from the one in which they first arose.

All of these outlets have received much attention in the past and about the first two nothing need be added here. But there is an interesting study of aggression in childhood which is worth considering in some detail for the light it throws on the general notion of alternative realization and on the third outlet in particular.

Margaret Manning made extended observations of a small group of children, first in nursery school and then in primary school a few years later.[6] The focus of interest was on the occurrence of hostile behaviour.

In order to determine which acts should count as hostile, Manning and her colleagues took the sensible course of noting how the acts were received. Protest, retaliation, crying and the like on the part of the child to whom the act was directed were the criteria for calling it 'hostile'. Then context was taken into consideration and three categories of hostile behaviour were established.

First, there was 'specific hostility', that is, hostility provoked by some immediate annoyance or frustration. Here the hostility was self-assertive, and it served as a means for achieving some goal – recapturing a toy, taking precedence over another child, insisting that the nursery rules should be observed.

Second, there was 'harassment'. For an act to count as harassment there had to be no obvious direct provocation; and in these cases there seemed likewise to be no extrinsic goal. The reward lay in the reaction of the victim. If, for instance, the aggressor destroyed something which another child had built, she would take care to see that the other child knew what had happened.

Hostile behaviour in these first two categories might be either physical or verbal. The third, however, was largely physical and consisted of extremely rough behaviour in the course of play – gripping round the throat, hurling to the ground. There might also be intimidation in the course of fantasy play – for instance, imprisoning victims against their will. Manning calls this 'game hostility'. (An extreme form of game hostility is presumably what led to the terrible trauma suffered by Bettelheim's patient – see p. 29.)

The children in Manning's study were sorted into four groups according to the relative proportions of the different types of hostility that they displayed. Two conclusions of interest then emerged. The first is that there was consistency in behaviour from nursery to primary school. Children who showed particular patterns of behaviour when first observed were still tending to behave in the same way four years later. The second finding is that relationships could be established between the behaviour patterns in the schools and circumstances in the children's lives at home.

The complex results that justify the second conclusion may be summarized as follows.

First, the children who specialized in game hostility came from families where, in Manning's words, 'all the members of the family seem to be negative to one another'. A further characteristic of these families was favouritism, and the 'games specialists' tended to be less favoured by mothers than their siblings. Manning suggests that, feeling disfavoured but wanting affection, they try to behave in ways outwardly friendly but actually hostile – for instance, they may threaten with a smiling face. Games thus afford them a splendid outlet.

Second, the children who specialized in harassment had home backgrounds of two kinds. Either they had one or more dominating siblings; or else they had dominating and overcontrolling mothers. Those with dominating siblings tended to harass other children at school by teasing them or, in general, to annoy by verbal means. Those with very strict mothers tended rather towards physical violence.

Manning says of the latter: 'These children are among the most hostile, most violent and most unfriendly children in the nursery.' With their mothers, by contrast, they were 'subdued and inhibited'. The mothers of this group were specially concerned about rules, appearance and good behaviour, and the rules were enforced strictly, often severely. One mother said: 'If a child has good manners, you can take him anywhere.' Concern, then, was with the child's outward demeanour and there seemed to be little interest in him for his own sake. Thus conversation between mother and child was very limited.

It was perhaps in this last respect that the difference between 'harassment specialists' and 'specific specialists' showed most clearly. The children whose aggression was largely directed to the achievement of some aim beyond itself – that is, who did not just attack for the sake of attacking – tended to come from homes where children and mothers talked freely and happily together, where conversation was lively and spontaneous, where many interests were shared. Discipline was significantly less severe. The mothers were friendly and not fussy. They were not necessarily specially warm and demonstrative, being outstanding not so much for hugs and kisses as for smiles and laughter.

There is good evidence here, then, to support the claim that human mentality is marked by a very special sort of tenacity. However, there is a danger of working with an oversimplified model of the processes we have been discussing. Once we recognize that undischarged impulses, especially those with very powerful emotion attending them, frequently refuse to 'lie down and die', it comes to seem reasonable that we should speak of them as finding other outlets – in dreams, fantasies, bodily symptoms or other overt behaviour. The risk in this way of talking is to assume that a fixed amount of energy accompanied the initial impulse and that when this is 'discharged' through some alternative 'channel', the impulse will *then* die.

This is the notion underlying the belief in catharsis – the notion that seeing an aggressive film, for instance, will arouse aggressive fantasies and will thus reduce the tendency to express aggression overtly since another outlet has thereby been provided and the energy has been 'spent'.

But the mind does not always work in this way. It is not an inert system of pathways, linking entrances and exits. Rather it is an active system for the interpretation of events and for the construction of meanings. Thus the 'hydraulic' model, as it is sometimes called, fails to fit because the energy of the original blocked impulse does not have to remain as a fixed quantity. The activity of the mind is entirely capable of amplifying it or, to take a more biological metaphor, cultivating it. Shylock, true to this fact, does not speak of discharging his hatred through his revenge. Rather he says:

> If I can catch him once upon the hip,
> I will feed fat the ancient grudge I bear him.[7]

There may sometimes be discharge and effective release. But often 'feeding fat' is a more appropriate description of what takes place.

So far we have been considering some of the emotional consequences of the human capacity to hold unrealized impulses in mind. But there are other consequences still to be explored. It is now time to take a different direction and look at the implications for rational thought.

At the heart of rationality lies the ability to make deductive inferences; and at the source of this lies the understanding of incompatibility – the recognition that one happening may preclude another. It is because we have this understanding that we can conceive of excluding possibilities and ending up with the 'only one remaining' or that which 'must be'.

This awareness can arise only in beings who are able to reflect on happenings as *possibilities realized*. So long as one thinks only about things as they are, one perceives no incompatibility. The world, directly observed, reveals none. Coexisting things are necessarily compatible. This stick is longer than that stick and there's an end on't. Inspection of the two yields nothing with which the fact conflicts. Again, this flower is red while that one is blue. There is no more to say, so long as we deal with a static world, seen in a cross-section of time.

Incompatibility pertains to a world of process, to a world continually coming into being – a world where if x comes into being y does not. Our first apprehension of this idea seems likely to arise as we start to participate in the process through our conscious purposes and intentional acts. (See chapter 4, in which there has already been some discussion of these notions.) The realization comes particularly when we are forced to recognize that impulses to action conflict with one another so that 'we can't have it both ways'. This is the source of our understanding of incompatibility and hence of our rationality. But it is because we have minds which later become able to represent the unrealized impulses in some enduring form that we come to think of reality as that which was possible and has actually happened.

Chapter 4 put the case for believing that we have the basic experience of conflict and choice at an early age. But this, of course, is only the beginning, the precondition. Before the emerging grasp of incompatibility can be used in the service of reasoning more must be achieved. It is necessary to go on to recognize that, while conflicting happenings cannot coexist, conflicting thoughts in no way preclude one another *as thoughts*. The mind can freely 'entertain' them, though the conflict may be experienced as demanding resolution. Only when this is apprehended – implicitly at first, no doubt – can there be a start to the business of putting ideas together, combining propositions and drawing conclusions. *The ability to combine propositions and draw conclusions is a prerequisite for the third mode.*

But when does it begin to appear? One difficulty immediately encountered is that reasoning of this kind cannot be recognized – even if it could occur – until it can find expression in language. And linguistic skills adequate to such a task take some time to develop – though less time perhaps than might be supposed.

Here is an example that we owe to a child called Sarah, aged two years ten months. Sarah was very naughty and her father spanked her. She reproached him indignantly as follows:

> 'You no spank Sarah!
> Sarah a lady.
> You no spank ladies!'[8]

This is an explicit syllogism. It has the form traditionally exemplified by:

> All men are mortal.
> Socrates is a man.
> Therefore Socrates is mortal.'

Thus:

> 'Ladies are not-to-be-spanked.
> Sarah is a lady.
> Therefore Sarah is not-to-be-spanked.'

The incompatibility that Sarah perceives (or at any rate chooses to invoke) is between the notion of being spanked and the notion of being a lady. Sarah is able to hold both in mind and use the incompatibility to draw and support a conclusion. She states the conclusion first and then justifies it, whereas the traditional presentation follows the opposite order. But the reasoning is the same, and

Sarah's order is the one that most of us would use, I think, in informal argument. So this little girl, not yet three years old, demonstrates that she is equipped with what we have to call rationality.

6

Two Varieties of the Construct Mode

> CALLUM: Is God everywhere?
> MOTHER: Yes, dear.
> CALLUM: Is he in this room?
> MOTHER: Yes, he is.
> CALLUM: Is he in my mug?
> MOTHER [*growing uneasy*]: Er – yes.
> CALLUM [*clapping his hand over his mug*]: Got him!
>
> (Callum was four years old at the time of the conversation.)

We have seen that the line mode achieves movement away from the present by shifting the locus of concern into the personal past and the personal future. But there is another kind of movement that we become able to make – a movement away from specific happenings towards a general concern with how things are.

In line-mode functioning perceptions and actions have fallen away, but the field of operation is provided by memories and forward projections. What has to be achieved beyond this is a movement of the locus of concern away from particular happenings in time. Instead of here/now or there/then the mind will next begin to concern itself with a locus conceived as somewhere/sometime or anywhere/anytime. Thus in the third mode we are no longer restricted to a consideration of episodes in our own experience – or even those we have heard about from others. We start to be actively and consciously concerned about the general nature of things.

This is a momentous development. However, when it first begins to happen, and for some time thereafter until the fourth mode appears on the scene, the mind still cannot function without a context of some familiar kind. The needed context is, by definition, no longer provided by the perception of specific events, by the memory of them or by the anticipation of them. So it has to be supplied by a deliberate constructive act of imagination. For this reason the third mode is called the construct mode.

The imagination is by no means without work to do in other modes, notably in the future-scanning activities of the line mode when we wonder: What might happen? Also the process of recalling the past has long been recognized as having its constructive aspects. However, the movement from specificity to generality makes new kinds of demand for context-building. Thus the name 'construct mode' seems appropriate.

The construct mode accounts for a large part of our mental activity. It contains much diversity, and there are subcategories that need to be recognized.

Like the line mode, the construct mode does not depend on direct perception or action. Point-mode activities may, of course, be combined with the construct mode as when Callum claps a hand over his mug; but they are not of its essence. This leaves us with the roles of thought and emotion to consider. And it is the variation of these roles that yields the subdivisions.

In some construct-mode activities the intention is to think unemotionally or dispassionately. But in other cases there is no such purpose: the aim of excluding emotion is absent, so that the concern with how things are takes the form of a free intermingling of thought and feeling. That is, the mind engages in some general inquiry and at the same time allows unrestrained emotional response to the topic.

Where thought and emotion are both present in this last way, on a more or less equal footing, we shall speak of the core construct mode. The earliest versions of the point mode and the line mode – that is, the ones we have been mainly concerned with so far – may also be called core modes in the same sense and are to be distinguished from other subcategories that develop later. For instance, there exists a kind of point-mode functioning, critically important for science, where the aim is to observe present happenings in a manner we call 'objective'.

In the case of the construct mode there are two subcategories, apart from the core one, which will need to be considered in some detail. These are called the intellectual construct mode and the value-sensing construct mode. Discussion of the core and of the intellectual varieties will follow in this chapter. Discussion of the value-sensing variety will come in chapter 9.

The intellectual construct mode is one in which a clear priority is given to thought. Notice, however, that a certain kind of emotion can properly attend – and normally does attend – the sort of inquiry we call dispassionate. After all, if we did not care whether we solved the problem or discovered the truth we would not put much effort into the enterprise. What is incompatible with the functioning of the intellectual construct mode is the kind of emotion that interferes with the inquiry, tending to bias the outcome – causing us, say, to reject some datum or argument because this conflicts with something else which we want to continue to maintain.

There are a number of causes of bias and distortion in human reasoning, but prior emotional commitment to a belief is undoubtedly a major one. To the extent that we do not rid ourselves of this kind of commitment we are functioning in the core construct mode. Sometimes there is no harm in this kind of functioning, but sometimes it is highly dangerous and undesirable. For example, Evans argues that it may have serious consequences when expert witnesses are called in court cases. As he says: 'there is widespread evidence in social psychology that people have mechanisms for maintaining beliefs and either avoiding or discrediting evidence which conflicts with those beliefs'.[1]

Unfortunately experts have these mechanisms available to them, just like other people, and unless the experts are highly self-aware, disciplined and scrupulous, then they may claim to be – and indeed believe themselves to be – reasoning rigorously when they are doing nothing of the kind. The risks are obvious.

So far, then, we have briefly considered two subcategories of the construct mode: one where thought has priority; and one where both thought and emotion have their due, and more or less equal, parts to play. We shall now look more closely at each of these in turn, starting with the latter kind.

As we grow up, we all develop beliefs about the nature of the universe, about our own social group, about other social groups – and about ourselves. Typically these beliefs are passionately held. They matter to us. They are loaded with value feelings.

It is possible to think dispassionately about such topics, as astronomers, physicists, sociologists and psychologists, for instance, contrive to do more or less successfully in their professional lives. But those of us who work in such disciplines are no more able than other people to live normal human lives without the emotional belief systems that constitute what we call our 'character' and our 'attitudes'. These are most intimately part of ourselves. We may contrive to set them aside while we think scientifically, though this is by no means easy. Indeed, the attempt to do it often fails, as we have seen, so that what is meant to be 'dispassionate' intellectual activity is nothing of the kind.

Also, those ways of functioning in which thought and emotion are closely interwoven tend to reassert themselves as soon as we take a rest from our professional pursuits. It is rather like driving a car which 'prefers' certain gears and slips back into them at the first opportunity. Thus the core construct mode is certainly not superseded as we grow older. Like the other modes, it changes; but it is not lost to us with the passage of time.

It seems likely that, at the very origin of this mode, there lies the first appearance of a reflective notion of the self. Some sense of one's self as an agent is probably present very early indeed, as we saw in chapter 3. But the construction of a self-concept, or self-image, is another matter. The development of the ability to think about, or conceptualize, one's self seems to have its origin around the age of eighteen months. About this time children start to show signs of knowing that when they look in a mirror it is themselves they see there. Younger children are apt to point at the mirror but children from eighteen months on do not ordinarily do so. And if the child's face is marked in some way – with a dab of rouge perhaps – clear evidence of the emergence of a new kind of

Development in Childhood

awareness can be obtained; for the older children show interest in the mark by touching or rubbing their own faces. They do not rub the mirror.[2]

Around the same age of eighteen months children also start to be able to talk about themselves. Often they use their names before they can use personal pronouns (I, me, you, etc.) correctly. That is, a child may say 'Kate want that' instead of 'I want that'. When children do start to use personal pronouns they may occasionally get muddled at first, saying 'you' when they mean 'I' or vice versa. This is scarcely surprising, since the child is addressed as 'you'; but it can have amusing consequences as when one child broke in upon his mother's tea party and said urgently: 'Mummy, your pants are coming down.'[3]

In spite of such confusions, which are usually quite short-lived if they occur at all, the verbal evidence leaves little doubt that by the end of the second year of life children can think about themselves and that they exercise this skill a great deal.

Those senses of self that are present during the early point-mode and line-mode activities need not depend on much understanding of how the self relates to the rest of the world. However, it is a different matter when the construction of a reflective self-concept is undertaken. For this is precisely a task through which we aim to know our place in the scheme of things – and our worth as part of that scheme. Thus it is unavoidably both conceptual and value-laden, cognitive and emotive.

It follows that the line mode and the core construct mode develop in close relationship after the advent of the latter, greatly influencing one another. Our interpretations of what happens to us, our plans and our projects are affected by the way we see ourselves in relation to others, and by some very general underlying assumptions about the nature of existence. Also the influence works just as strongly in the other direction: there is a constant and powerful interaction. Nevertheless, there are still two modes, distinguished by the locus of concern that is dominant at any given time.

It is obvious that, in attempting to build a self-concept and a system of general notions about the world he lives in, a young child will be much influenced by the concepts and values of those around him: first by what is said and done in his immediate family, then by the widening range of his social encounters. Children, however, do not merely wait for these experiences to provide data. They certainly use the data that just happen to come their way, but, given any kind of encouragement, they soon initiate searches of their own: they ask questions. Some children between the ages of three and six, or thereabouts, do this endlessly.

Of course, the seriousness of the search may vary. Callum's inquiry as to the omnipresence of God has a playful, teasing ring. He is not too worried. He is having fun. On the other hand, there may be anxiety in some of the questioning that occurs. Gareth Matthews points this out when he quotes John who, at the age of six, reflected that his toys, his arms, his head are all his and wondered: 'Which part of me is really me?'[4]

Less profound questions which are still aimed at general under-standing are also very common. Barbara Tizard and Martin Hughes provide us with examples from a wide range of everyday topics: for instance, how the size of a person is related to the age of a person; what it is to be left-handed; why Mum gives money to the window-cleaner instead of receiving it from him; why most of the rosehip syrup has to be kept for the baby (though the child who asks the question wants it and is jealous of the baby); why Mum (who has a lot of housework to do) can't spend much time in play; and so on – and on.[5]

These topics vary greatly in emotional power. But they are being pursued through a mode of mental functioning where there is no effort to keep emotion out, no sense that it should be kept out. Whenever it arises, it is freely allowed in. And it is appropriate for this to happen. The inquiries are not – and are not meant to be – detached and dispassionate.

It is important to consider how far we are 'shaped' in all those matters by the culture in which we are reared. What choice have we in the end – or in the beginning? This question is one that tends to elicit opinions with a good deal of emotion in them.

As to the beginning – that is, the origins of the core construct mode – some have argued that young children have very little choice at all except to see themselves and the social world through the eyes of those who are in charge of them. For instance, as Peter Berger and Thomas Luckmann put it, the adults 'set the rules of the game' and the child's contribution amounts only to determining whether to play 'with enthusiasm or with sullen resistance'. Berger and Luckmann allow that the child is 'not simply passive'. Nevertheless, according to them:

his internalization of their [the care-givers'] particular reality is quasi inevitable. The child does not internalize the world of his significant others as one of many possible worlds. He internalizes it as *the* world, the only existent and only conceivable world, the world *tout court*.[6]

There is at first sight a certain compelling quality to this argument. For how could a very young child conceive of other possibilities than those offered to him? And clearly children do accept a great deal 'on trust', as it were. But the claim of inevitability – even moderated by 'quasi' – seems less convincing when one takes account of the evidence discussed earlier about young children's cognitive capacities; and it seems much less convincing when one pays attention to the things many children actually say. Their conversations are so full of doubts and wonderings – and, often, of an ability to 'take on' adults and confound them.

Here is a relevant example, taken from an occasion when a father was trying to test the intelligence of his three-year-old son. The first question was about gender identity. The exchange went as follows:

FATHER: Stephen, are you a little boy or a little girl?
STEPHEN: I'm a little doggie.
FATHER: Come on now, Stephen! Be sensible. Are you a little boy or a little girl?
STEPHEN: Gr-rrr! Woof!

The attempt at testing ended in helpless laughter. Who may be said to have set the rules of the game?

The book by Matthews, cited earlier, is a rich source of illustrations like this one: illustrations of children's ability to consider
what they are told instead of just accepting it, and also to play
with possibilities.[7] But the presence of adults who take part in
dialogue willingly, or, better still, enthusiastically – as Matthews
did with his own son – is clearly helpful. Children can be
encouraged to pursue their early careers as independent thinkers –
or they can be discouraged. On the other hand, some are no doubt
easier to discourage than others.

So children find out about the way things are by questioning –
but not only by questioning. They also watch and listen, and are
offered information or instruction. They look at television, they
play, they argue – in due course they read.

In these ways they gradually construct highly complex belief
systems which become essential in their lives. To give these the
name 'systems' is not to imply that they are always coherent or
internally consistent. But the term 'system' is justified, because we
are not usually content with beliefs that are isolated. We do our
best to make them 'hang together'.

The beliefs thus arrived at relate to the self-image – how brave
one is or how cowardly, how clever or how 'dumb', how loved
and how lovable. They also relate to the nature of the world,
particularly the social world. Notice that nowadays, when understanding of the physical world is at issue, we strongly encourage
children, as they grow older, to try to use the relatively unemotive
intellectual modes. But it has not always been so. We shall see later
that this seems to have happened fairly recently, which means that
the scope of the core construct mode is now restricted, compared
with what it used to be. However, the range is still very wide.
Most of our opinions on matters political, economic or religious,
as well as our judgements about close interpersonal relations, are
the outcome of its functioning.

There will be much more to say later about the core construct
mode. We come now to the early development of the intellectual
version of this mode. And, perhaps surprisingly, its advent does

not seem to be very long delayed after the onset of the core category. One can, of course, never determine these matters with any degree of precision. However, it is possible to find evidence of what we have to call intellectual activity in many children from around the age of three years. To illustrate this I turn to the work of Martin Hughes.[8]

It is well known that children of three and four can often count sets of objects shown to them, so long as the numbers are quite small. What Hughes found is that, with the same proviso, they can also answer number questions about hypothetical situations. Questions like: 'If there were two girls in a shop and one more went in, how many girls would be in the shop now?' were answered correctly by 62 per cent of the children taking part in his study.

Now clearly the kind of thinking that these problems call for reaches beyond the point or line modes. It is not about anything here and now, and it is not about a past or a future event either. The girls in the shop are not specific girls in a specific shop on a specific day, even if they are imagined quite vividly as the problem is being tackled. They are any girls in any shop at any time. Imagining other girls in other shops would yield the same answer. Thus generality has been attained.

However, another crucially important finding emerged from Hughes's work – a finding that helps to clarify the nature of the intellectual construct mode by showing how it differs from the other intellectual mode that will follow after it. (This is called the intellectual transcendent mode and is discussed in chapter 8.) There was something that most of these same children could not do at all – something that seems to an adult closely similar and little, if at all, more taxing. They could not add one and two. That is to say, they could add one brick and two bricks, or one child and two children or whatever – even without actual bricks and children to look at and count – but they could not add numbers.

There was a boy called Ram, aged four years seven months, a bright and competent child, who made his problem explicit in the following interview with Hughes:

M.H.: What is three and one more? How many is three and one more?

RAM: Three and what? One what? Letter? I mean, number?

[*He was referring presumably to magnetic numerals used in an earlier game.*]

M.H.: How many is three and one more?

RAM: One more what?

M.H.: Just one more, you know.

RAM [*disgruntled*]: I don't know.

Ram and most of the other three- and four-year-olds in the group were able to accept certain questions that had no bearing on their own immediate concerns or on the course of their own lives, and to answer these readily. On request they were able, as they sat in a room with the experimenter, to turn their minds to hypothetical situations and reason about them. They could think of a shop with two children in it; they could think of the departure of one of these or the entry of a third child; and they could compute the consequence of the change. This is no small achievement. Yet they could not think of one and then add two. Why not? What is it that Ram so emphatically does not know? Why can he not function without an answer to the question: 'One more what?'

Hughes is inclined to view the difficulty as one of learning the language of mathematics and becoming skilled at 'translating' into or out of it. There is good sense in this and the notion has applications to some of the discoveries that Hughes made when he worked with older children. However, I do not believe that Ram's fundamental problem is one of learning a new way of talking; or at least, if we are to think of it that way, we must ask why this new way seems to present such particular difficulty to him and to many others.

A. N. Whitehead, writing on mathematics and the history of human thought, has some relevant things to say.[9] He conjectures that pure mathematics began when some unknown person long ago moved from thinking of seven fishes or seven days to thinking of seven. In other words, Whitehead is saying that this unknown genius overcame what is precisely Ram's problem and conceived of a pure number, so becoming able to think of 'one more' without having to know 'one more what'.

Many of us, lacking Whitehead's insight into the magnitude of the achievement, are surprised to learn how hard this is for young children to do. And almost everyone supposes at first that all the children need is to be encouraged to think about many concrete examples – various sets of apples, zebras, balloons, and so on – from which they will then learn to 'abstract' the numbers alone. I thought so myself when I first heard of the finding. But it proves not to be so easy, as Hughes reports.

Like Hughes I do not find it satisfactory to see Ram's problem as arising from lack of skill in 'abstracting'. Certainly number on its own is highly abstract. Yet to emphasize this, as even Whitehead does, is to miss the main point – to fail to see what is really so hard. For all thinking involves abstraction. This has often been pointed out but it is still often overlooked. To recognize that a particular poodle is a dog and that a particular Great Dane is one also is to abstract qualities which the two have in common. Yet children do this readily and very early, showing great skill as soon as they begin to talk; and the skill is certainly present before then. (Some of the relevant evidence has already been discussed in chapter 3.) After all, a very young child can recognize a blob of dark wool on the face of a teddy bear as a nose. Some very abstract attributes must be picked up in that act of categorizing.

What is involved in the mind's movement from 'seven fishes' to 'seven' is abstraction indeed, but it is more: it is a dramatic decontextualization. In the contexts of our ordinary life we have to deal with quantities of fishes but we never encounter seven.[10]

A pure number resists embedding in any human context. You can catch fishes and eat them. You can love dogs and fear them. You can see them and touch them. You very readily can have purposes that involve them. And so it is true that being concrete in the direct sense of being tangible has to do with the difference. But the underlying distinction – the one that is of the most profound psychological importance – is ease of relation to the personal life.

In general, human beings learn early and readily to cope with things that bear directly on their purposes, their hopes and their fears, even though sometimes this coping entails a great deal of conceptual abstraction. And when some measure of reflection on the general nature of things becomes possible, ease of relation to things personally known does not cease to be relevant. This is true not only of the core construct mode but of the intellectual construct mode as well. In the kind of intellectual problem that depends on the imaginative construction of an embedding context – or scenario, as one might call it – direct experience remains helpful. The constructive task is less onerous if what we need to imagine is close to what we know well, if the appropriate context is one in which we feel 'at home'.

This is scarcely surprising. Less obvious, perhaps, is how easy the constructive task can often seem once the general nature of the required setting is grasped, which is one reason why the nature of thinking in the intellectual construct mode has not always been evident.

Consider the following case taken from the work of Bransford and McCarrell, who were studying certain aspects of the comprehension of language and who used a number of sentences which their adult subjects were apt to find perplexing.[11] An example is: 'The haystack was important because the cloth ripped.' This is a perfectly meaningful sentence. (To see that this is so, contrast it with the example beloved of psycholinguists: 'Colourless green ideas sleep furiously.') However, most people are bothered by what seems a lack of sense in it until they are told that it is 'about' a parachute jump. This information instantly brings with it what Roger Brown has called, in a famous phrase, the 'click of comprehension'.

Literate adults, as Bransford's subjects were, are quite used to interpreting sentences in isolation and usually do not find it hard to make sense of them. So why did this sentence – and some others like it – give such difficulty? What Bransford and McCarrell's results strongly suggest is that normally, when we interpret language that does not refer to actual episodes of which we have knowledge, we supply suitable contexts without noticing what we are doing: we embed the language by imagining settings in which the words might reasonably have been uttered. On this view the sentences used by Bransford and McCarrell proved baffling because the needed context was quite unusual and there was no clue to suggest it. Thus the click of comprehension at first failed to come.

On the other hand, once the clue was given all was immediately clear. Yet most people have never come down by parachute. How then are we to account for the rapid, easy clarification?

The point is that, although we may never have jumped from a plane, we have a very well-established, personal knowledge of jumping and falling; we can easily see how a haystack can serve to limit the harm done by a fall; and we understand the significance of the integrity of a parachute in relation to the purpose of escaping injury. The interesting thing is the speed and ease with which, given the right conditions, these components of understanding come together in the imaginative activity we call 'making sense'. The first and second of the components are acquired early and universally: they are part of human sense (see p. 95). The third is certainly part of the shared knowledge of all older children and adults in our kind of society. This background of shared knowledge is crucial for successful communication.

The case of the parachute jump provides a good opportunity for illustrating the difference between the point mode and the construct mode. Suppose one were standing in a field talking to someone who had just made such a jump. Then point-mode functioning would yield the click of comprehension just as soon as a sentence linking torn cloth to the importance of a haystack was uttered. Given that the relevant events are not currently taking place but that one is able to imagine the setting without difficulty, construct-mode functioning quickly and smoothly takes over instead. Of course, we do not have to envisage the scene with any of the vividness of detail that an author writing on the subject might try to evoke. This is not essential for the making of sense; but what is involved is an imaginative activity all the same.

Except when the context is quite unusual and when clues as to its nature are sparse, this kind of sense-making is not generally experienced as deliberate. However, the ability to bring context-building under deliberate control starts to show itself from around the age of three, and it is very important for the cognitive developments that lie ahead.

Mental functioning in the first two or three years is marked by a great deal of spontaneity. Thoughts, emotions, memories and anticipations arise out of ongoing events. Young children do not in general 'turn their minds' to particular topics or 'hold their minds' on one thing for very long. The following example, when compared with the kind of control shown by Hughes's three- and four-year-old subjects, will illustrate what I mean.

Laura was a member of a small group of two-year-olds with whom I worked regularly over a period of some months. Part of my purpose was to look for evidence of the beginnings of numerical understanding, but Laura had shown none of this at all: every attempt to elicit it had failed.

One day, when Laura was two years eight months old, she asked to have a story read to her and the choice fell on a book about Desmond the dinosaur and his holiday visit to Loch Ness. It will come as no surprise to the adult reader that when Desmond reaches Loch Ness he meets the monster, but it came as a great surprise to Laura. We turned a page – and there was a picture of Desmond on the shore looking at another creature whose head and neck emerged from the water. 'That's the monster,' I said, pointing.

Now the dinosaur had not previously been called a monster, but the illustrator had made the two creatures look very much alike. Laura instantly saw the resemblance and was excited by it. She pointed to the one, saying: 'Monster!' and to the other, saying: 'Monster!' again. Then she looked at me with shining eyes and said: 'Two monsters!'

My interest in Laura's number knowledge was greater than my interest in the monsters so I tried to seize the opportunity. 'Laura,' I asked, 'how many hands have you got?' This question evoked the answer: 'One, two, six, seven.' It also evoked a very strange little sideways glance at me. Then I had to laugh at myself, for I realized how inappropriate my question had been. Laura and I were looking at a picture of monsters. This was the focus of concern and it was there before us, directly perceived. It was what we were talking about. My question, however, had called for the deliberate turning of the mind to an abruptly different topic – one that gave rise to no feeling of spontaneous interest or excitement. So Laura could make nothing of it. I believe that she was still incapable of dealing with the kind of demand for deliberate control that it placed upon her. Her spontaneous exclamation about the monsters showed clearly that she had some knowledge of number (at least up to two) but she could bring this into play only in certain circumstances. She needed a context that was appropriate and sustaining. She was not yet able to construct one 'to order'.

It is essential to recognize that the context which sustained Laura's thoughts was not just the physical object – the picture-book – nor yet just my words as I told the story. It included, most importantly, her own ongoing thoughts and feelings – her surprise at the discovery that Desmond could be regarded as a monster and her interest in this finding. Also it included her desire to impart the discovery to me. She spoke of 'two monsters' when the intention to communicate rose spontaneously and strongly in her. But she failed to marshal her knowledge of number in order to answer a question put to her by someone else when an appropriate context was lacking.

At about the same time as the intellectual construct mode first manifests itself children start also to be able to accept from other people questions that are 'abrupt' in the sense just illustrated. I am unsure at the moment how these two capacities are related to one another. Perhaps it is the new capacity for control and deliberate direction of the mind that makes possible the start of what we call intellectual activity. Or perhaps the two are more closely interwoven. Perhaps – as Piaget liked to say when there was really no way to separate two aspects of development – we have to think of them as *solidaires*.

I should emphasize at this point that, in setting the start of intellectual activity as late as age three or thereabouts, I am distinguishing 'intellectual' from 'intelligent'. Much intelligent behaviour is clearly observed before then.

Construct-mode functioning of the intellectual kind is favoured in its early manifestations by certain circumstances. It helps greatly if the topic is not of any special personal concern to the child so that it is unlikely to give rise to the kind of emotion that would have to be resisted. Where the thinking consists in trying to solve a problem set by another person it helps, too, if the child is not strongly interested in anything else at the time the question is asked, and if the questioner is known and trusted. And it matters that the effort of imagination involved in the provision of an embedding context should not be too great, for if this part of the task is too hard the whole enterprise fails.

Although children of three are only at the start of their attempts to investigate the general nature of things, they have already acquired a great deal of basic knowledge in the course of their point-mode and line-mode activities.[12] In these activities, concern has been with specific happenings and how to cope with them. But the experience gained provides the essential ground for what is to follow, for it has yielded an easy familiarity with many common social and physical events and a basic recognition of other minds having purposes and feelings like one's own.[13]

Together these constitute what we may call human sense — a vitally important base of knowledge and understanding which all normally developing human beings hold in common and which we take for granted through the rest of life. Within its limits we know what we are about. We are at home. And generally we feel that we can function surely, confidently — so confidently that we say we do it 'without thinking'. We know that balls will roll while cubes will not, that some things will prove heavier to lift than others, that liquid will spill if a cup is turned upside-down. Similarly we understand offers of help and requests for help. We understand playful teasing and serious threat. We understand flight and pursuit, concealment and search, and so on.

The kinds of intellectual construct-mode problem that children can first begin to deal with are those which call mainly on this fundamental knowledge; though it has to be handled in a new way. However, there is an important further consideration. It is not just the case that much of this common base is knowledge about human intentions. It is also true to say that the whole of it has been developed, and is ordinarily used, in the course of purposive human activities. What, then, is the relevance of intention when, in place of activity arising spontaneously from individual or shared goals, there comes an abrupt question to be tackled on its own, 'to order'?

We have at once to make a distinction between children's attempts to understand the intentions of the person who presents a problem to them and their attempts to understand any intentions which may be at play within the problem itself. We shall leave the first of these aside for the time being and come back to it later. As to the second, its significance will become clearer if we consider an example.

Martin Hughes devised a problem in which the child's task is to hide a 'boy' – a wooden doll a few inches high – in some position behind a wall (a toy wall on a board) so that a 'policeman' set at a fixed point in the array cannot see him. The accompanying story is that the boy has been very naughty and the policeman is looking for him. So where can he hide?[14]

Now according to Piaget three-year-old children are egocentric in the sense that they cannot mentally adopt a point of view other than their own. If this were correct, then the children in Hughes's study should have placed the boy where they themselves could not see him instead of worrying about the line of sight of the policeman. They should have assumed that the policeman's point of view was the same as their own. But children of three were able to tackle the policeman problem with spectacularly high rates of success.

I believe that this is because the problem was set so squarely in the realm of human sense that the imagination could operate with ease. The children could quickly comprehend the intention to escape, the intention to pursue, the significance of an available wall and of the line of sight of the pursuer.

To this it might be objected that the child's imagination was actually not called on at all – that the toys were provided and that the setting was virtually given on the table-top. But this is not so. The toys in such a task are merely supports for the activity of imaginative embedding. The children are called on to interpret the language of the instructions at a time when they are not themselves in a real-life setting of an appropriate kind. They have no activated motives for running away or for chasing. They are not in the problem situation, they must enter into it in their minds. However, in the case of the policeman problem the intentions at play within the problem – and the spatial relations involved – are ones that children very well understand so that 'entering into it' is quite easy. Often in problems that look equally simple to an adult it is not so.

A study by Sally Collie helps to make plain the imaginative nature of the embedding activity in such a case. Collie's subjects were forty nursery-school children, aged between three years six months and five years. The main point of the research was to vary Hughes's task by comparing a 'sensible' condition with a 'nonsense' one. For the former Collie chose a mouse hiding from a cat, for the latter a stone hiding from a tree. All the children were given both problems, but sometimes 'sense' came first and sometimes 'nonsense'.[15]

The different orders of presentation had a marked effect. If the children encountered the sensible task first, they did well, just like the children in Hughes's study. What is striking is that, having helped a mouse to hide from a cat, they could help a stone to hide from a tree without signs of puzzlement and with equivalent levels of success. However, they managed significantly less well if the trees and stones came first. For in this case the initial imaginative demands were greater: the appropriate context was harder to construct. Yet once they had taken the first step with the easy problem these three- and four-year-old children were already capable of imaginative extension to things never directly experienced.

Recent research is starting to throw new light on the early

development of imaginative powers. It is well established that when children are asked to reason from premises contradicting what they know to be true they often do this very poorly. So indeed do many adults, particularly those with no formal schooling. But Dias and Harris have now shown that, if steps are taken to encourage treatment of the premises as fictitious or in some way free from the constraints of reality, then even children as young as four will often reason correctly from them. Of the methods used to achieve this result, direct invitation to 'make a picture in your head' proved the most effective. This is a very important finding.[16]

The claim I have been making is that hiding tasks of the kind used by Hughes rest upon a well-developed understanding of intentions and of how these interact. If a child has this understanding, then the child has knowledge – knowledge in some form – of human minds and how they function. That there is such knowledge shows itself in much early spontaneous behaviour, as we have seen; and it shows itself – later, of course, but still generally by the age of three – in spontaneous language use. During the course of the third year children commonly start to talk about mental states, using verbs like 'know', 'think', 'hope' or 'pretend'. They do this freely, appropriately and with skill. Yet recent studies have revealed what seem to be marked deficits in the typical three-year-old's understanding of such matters.[17]

Suppose, for instance, that a child is asked to consider the following kind of situation, enacted with two dolls, John and Peter. John puts a toy in a box (box A) and goes away. While John is away, Peter moves the toy to box B. Then John returns. Now the question for the child to answer is: where will John first look for his toy? What transpires is that between age three and age four a marked change takes place in the replies given. Three-year-olds for the most part predict that John will look in box B while four-year-olds generally choose box A. That is, four-year-olds are able to take account of the fact that John did not observe the move from A to B and so may be assumed to believe, falsely, that the toy is still where he left it. But three-year-olds generally seem not to be able to work this out. They give as their answer the place where they know the toy really is. That is, they predict that John will look in box B.

Several variants of this kind of task have been developed. For example, a child may be shown a box of the sweets known as Smarties (or M & M's) and be asked what is inside. In the great majority of cases the answer will, of course, express the normal expectation: 'Smarties'. The child is next shown that the box actually contains not Smarties but, say, a pencil, and is then asked to predict what another child will think to be in the box when first shown it.

From studies of this sort it has emerged repeatedly that four-year-olds take account of what knowledge another person is likely to possess in this kind of situation whereas three-year-olds usually do not. The typical three-year-old, knowing there is a pencil in the box, seems to think that someone else who has not looked inside will know this too.

Piaget would certainly have taken this to be yet another instance of the pervasive egocentrism of children of this age. He argued that children treat their own 'points of view' – literal or metaphorical – as a kind of false absolute, believing the world to be just as they see it, or as they know it, not recognizing that others might have different perceptions, different knowledge. Piaget's claim was a strong one, namely that children think like this until a certain stage is reached because they can do no other. However, there is now much evidence that the strong claim at least is wrong.

The hiding tasks we have been considering are by themselves enough to show this conclusively.[18] So if children do behave egocentrically in certain circumstances we still have to explain what there is about these circumstances that leads them to do so.

Compare the questions about hiding and searching with the questions about belief. Both require the child to make predictions which override his own perception or knowledge. He can see the boy who is hiding. Likewise he knows which box contains the toy, or that there is a pencil in the box of Smarties, or whatever.

But a correct answer requires him to work out what someone else would see or what someone else would think. He must calculate where someone else 'stands' – perceptually in the one case, cognitively in the other.

Now we have seen that roughly a year of development appears to separate these two achievements; and at first sight it does not seem surprising that perception should be easier to predict than mental states. However, we should be wary about drawing this general conclusion too hastily. In research of this kind much depends on the precise demands that a given task makes on the child's capacity for imaginative construction. Much depends too on the exact wording of the questions asked. For example, Wimmer, Hogrefe and Perner gave children aged between three and five a task that seems very straightforward.[19] Pairs of children sat opposite one another at a table on which lay a box. Sometimes one child was shown what the box contained, sometimes they both were. One child was then asked if he knew what was in the box and also if the other child knew it. The most frequent error, very common among the three-year-olds, was that child A might see child B looking in the box yet deny that child B would then know what it contained.

Wimmer used this and other related evidence as the basis for claiming that children at this stage do not understand what he calls the 'informational conditions' for knowledge and belief.[20] That is, while three-year-olds certainly have much knowledge and many beliefs, they are not aware of the origins or sources of these mental states. More specifically, they do not realize that they – and other people – know because they perceive things or are told things or make inferences.

Given the evidence Wimmer quoted, these arguments seem persuasive. Yet it turns out that a change in the way the questions are put to the children can make a very big difference. The form of question used by Wimmer and his colleagues was: 'Does x know what is in the box or does he not know that?' (The original questions were in German.) However, in a later experiment Chris Pratt and Peter Bryant tried the effect of using a shorter, more familiar form, dropping the second part of what they call the 'double-barrelled' question and asking only: 'Does x know what is in the box?'[21] This made the task considerably easier. Significantly more three-year-olds were now able to answer correctly.

Kate Sullivan and Ellen Winner have also looked at the effects of varying the way in which this same task is set.[22] They spoke explicitly to the children about the fact that beliefs can be wrong; they explained that this may be due to lack of information; and they presented the task as a game in which the purpose was to play a trick on someone.

Three groups of children took part: the 'young', the 'middle-aged' and the 'old'. The 'young' were aged between two years eleven months and three years three months; the 'middle-aged' were in the range three years four months to three years seven months; and the 'old' were between three years eight months and four years. It turned out that members of the old group were helped by all three variants, but for the middle-aged and the young there was no effect. That is, it was only the children who were approaching the age of four who seemed able to use the help offered.

However, Sullivan and Winner make a significant observation. They note that, when playing the game of tricking someone, children in the younger groups sometimes spontaneously showed that they understood the nature of the trick quite well. They would make remarks like: 'He's never going to know what's in here!' or: 'No one will know that we put pencils in here!' And these remarks were made 'gleefully'. Yet the same children who made them went on to answer 'yes' to the question: 'When x [another child] sees this bag all closed up will he know what's really in the bag, yes or no?'

This seems perplexing until one recognizes it as another instance of the regularly occurring discrepancy between spontaneous behaviour arising in the pursuit of an intention (in this case to play a trick on someone) and behaviour that entails a deliberate 'turning of the mind' in a direction towards which there is no immediately prior inclination.

Two important points of principle need now to be stated. The first is that experimental studies of the early functioning of the construct mode are not easy to conduct because the outcomes are sensitive to procedural differences that are apt to seem trivial to adults and consequently are easily overlooked. (There will be some more discussion of this in the next chapter.)

The second point is that, once the conscious search for general understanding has begun, it tends to be widespread. The same kind of reflective activity that is turned upon physical and social happenings is also turned upon mental happenings. Thus we should expect gain from attempts to relate the emerging evidence about children's 'theories of mind' to what is known about the development of thinking in general.

The problems that a child can solve when the intellectual construct mode starts to function around age three or four seem – and indeed are – simple. For all that, they lie at the beginning of a line of development which extends onwards to science, mathematics and logic.

As we have seen, thought of this kind is distinguished by one special characteristic: it is – or at least aspires to be – dispassionate. To succeed in the aspiration it must break free from entanglement with all goals other than those intrinsic to the thinking itself. That is, its only purpose must be to achieve some new insight or clearer understanding.

Emotions always arise in connection with some sense of importance. If the only important thing is to solve the problem, then success or failure in solving the problem will be the only source of emotion. The intellectual advantage is evident: there are then no extraneous emotions to interfere with the thought processes and affect their outcome. All of this is well recognized. It is the reason why dispassionate thought is valued.

We may say, then, that intellectual construct-mode activity is marked by a new kind of separation of thought from emotion. Yet it would be a grave mistake to suppose that the separation is complete. Einstein talks of his 'passionate curiosity'; and the making of any genuinely new discovery is normally a powerfully emotive experience. Nevertheless, with the advent of the intellectual construct mode a further step has been taken by way of separation of the components of mental activity. We have arrived at a mode of functioning where the locus of concern is anywhere – or at least in no specific place and time – and where thought has achieved a new kind of separation and independence. We are also on the way towards the fourth mode, called the transcendent mode. Further possibilities lie open before us.

An extremely important question now arises: does there exist, or could there ever exist, a parallel development for the emotions? Or, on the other hand, does emotional development reach a necessary upper limit in the core construct mode? If there is such a limit – if something stops it there – then this leaves a marked, and awkward, asymmetry between the intellect and the emotions.

It is often said that humanity currently suffers from a dangerous imbalance, attributable to progress in science and technology that has not been matched by comparable steps forward in the development of morals and emotions. If this is so, we need to understand how it has happened and whether we can do anything about it. What has made the intellect so powerful? Is it really the case that only thought can move onwards independently, becoming further disembedded?

These questions demand to be asked at this stage in the argument; but, having asked them, I shall lay them aside. They will be raised again in chapter 9. Intervening chapters will take up the theme of later developments in the intellectual construct mode and beyond. Before these developments can be discussed, however, some prior reflection is necessary on human capacities of the kind called 'representational' and then on certain questions specific to the learning of the mother tongue.

7

Language in Relation to the Modes

The notion that we are able to 'call things to mind' – to present or 'represent' them to ourselves for consideration – has figured in much of the previous discussion. It is now time for some reflection on this notion itself and on the nature and growth of our representational resources.

We commonly talk of 'envisaging' the things we think about and visual images are indeed one means available to us. But we have other resources too, including images in other sensory modalities and, most importantly, language.

Representations may be engendered privately, inwardly. They may also be external. Thus as well as visual images in the mind we can construct drawings. As for language, it has more than one external form, for it may exist publicly as speech, or as sign language, or as writing. But it also has a private mental form: we may think with words that we do not utter.

The claim has sometimes been made that the words with which we think *are* the thoughts – that language and thought are one and the same. I do not believe that this can be sustained. It is like saying that when we reach out with our hands the hands are the reaching.

Thinking is an activity in which we engage. We need our representational resources to make the activity effective, but the resources are varied, and they are not the same as the thought that they sustain.

A distinction must be drawn between cases where there are conventional agreements about what representations 'stand for' and cases where there are not. Language falls into the first category, private mental imagery into the second.

Where there is a system with agreed rules, the use of the system entails knowledge, implicit or explicit, of how the system works. But there is more than this to the use of language, for it also entails (among other things) judging the degree of fit between parts of the system and parts of the world. It entails being able to compare a representation with that which it is supposed to represent in order to determine if it really does so, according to the accepted rules.

Premack says that this last skill *is* 'representational capacity'; and, in the course of comparing human and animal skills, he asks whether a bee possesses it. A bee can use the dancing of another bee as a source of information about where food is to be found. But for Premack – and I agree with him – the crucial question is: could a bee, if it had independent knowledge of the food's location, judge whether a given dance was an accurate representation? In other words, could it tell whether the dance was *true*? Could it recognize error?[1]

Also, would it be possible to use variants of the dance to interrogate the bee, asking: 'Is the food there? Or there?' If not, then the bee does not use its dance in the way human beings use language.

It is normally during the second year of a child's life that the processes of finding out about language really get going. There are individual differences, of course, but in general, while little linguistic knowledge is evident at the age of one year, by the age of two most children appear to understand much that is said to them and can also speak with what amounts already to considerable skill. Thereafter progress is usually very rapid so that, before the age of three, children can use many words and many grammatical constructions. To give a few examples taken at random from records of spontaneous speech,[2] Linda, at age two years ten months, says: 'I haven't got a panda now because I changed my panda to a mouse.' And when asked if she wants help in taking beads off a string she replies: 'No, I can take them off by myself.' Janet, at age two years eight months, says as she plays with building blocks: 'That's the chimney going out and that's the piece that goes in there.' Stephen, age two years ten months, says: 'I had that game last time.' Also he asks questions like: 'Where does that one go?' and makes remarks like: 'I was trying to put it on there.'

How this level of skill is achieved so quickly is a puzzle that has occupied many minds, especially during the last few decades of active research in developmental psycholinguistics. Yet the achievement remains essentially mysterious. We do not know in any detail how it is done.

However, although the mystery abides, one thing seems clear: language in its origins is functionally intertwined with the rest of mental life. When it begins towards the end of the first year or early in the second, it is a new phenomenon undoubtedly but it makes its appearance as a component of typical point-mode activity. It is used in the furtherance of those other activities which, from moment to moment, occupy the child's mind. And it can be used at first in no other way.

Around the age of nine or ten months (that famous age) there occurs a development that is a precursor of language: children become able to encompass a person and an object within a single field of attention.

In earlier months babies concentrate exclusively on person or object, one at a time. But between eight and twelve months, as Hubley and Trevarthen have shown, it happens with increasing frequency that child, adult and object form a triad, so that communication *about* something can begin.[3]

This is Elizabeth Bates's first 'Moment' in language development – the moment when, according to her, the child starts to communicate intentionally. It is at the same time the moment when communication becomes referential.[4]

Bates takes the example of a child who is reaching for an object while an adult is nearby. Before the age of eight or nine months the baby typically will not turn towards the adult while reaching even if she fails to get the object and starts to cry in frustration; though she may, of course, turn to the adult for comfort, having given up. But around nine months the behaviour changes.

First, the baby now starts to look to and fro, from the desired object to the adult and back again, often making sounds as she does so. And soon she begins to send what are clearly signals, signals which she modifies in accordance with the way the adult responds. These signals gradually become 'ritualized', 'turning into shorter, more regular sounds that shift in volume depending on the action of the adult'. The baby in effect is asking for help.

Bates gives a detailed illustration, which is worth quoting in full:

Marta is unable to open a small purse, and places it in front of her father's hand (which is resting on the floor). F does nothing, so M puts the purse in his hand and utters a series of small sounds, looking at F. F still does not react and M insists, pointing to the purse and whining. F asks: 'What do I have to do?' M again points to the purse, looks at F and makes a series of small sounds. Finally, F touches the purse clasp and simultaneously asks: 'Should I open it?' M nods sharply.[5]

Are we to say that M understands her father's question: 'Should I open it?' Let us say rather that she interprets it correctly in its highly supportive context – or perhaps, better still, that she correctly interprets her father's behaviour, of which speech is one component. In this way the learning of language begins.

Thereafter the child soon goes on to make the great discovery that adults can teach by telling. Once this is understood children normally become highly responsive to verbal instruction, seeming to look for it and welcome it. And a very important step has been taken.

It is reasonable to expect that language, beginning in the point mode, will then be used in the line mode before it is available for construct-mode functioning, and this is indeed the case. Typically it is from the second birthday onwards that children begin to talk about things which have happened or which they expect to happen. For some time such topics remain rare by comparison with those that are about ongoing events. Nevertheless, even before the age of two children sometimes manage to use language for line-mode purposes. Here is an example.

Beth, at the age of twenty-two months, appeared to be alert and intelligent but had been slow in speaking. Until the two episodes that are about to be reported she had used her vocabulary of around fifty words mainly to label things seen around her or in picture-books. Thus she had spoken in the point mode and with single words, not phrases.

Then one day, during her mother's absence, the dog behaved very badly and was put outside. The following conversation (noted verbatim immediately afterwards) took place when the mother returned:

BETH [*highly excited*]: Doggie – doggie out. Out doggie.
MOTHER: The doggie – what's the doggie done? Where is she?
BETH: Doggie out. Doggie doggie out.
MOTHER: Did the doggie go out? Good God, it's a two-word utterance!

It might be possible to argue that Beth was simply pointing out to her mother the dog's absence – saying, in effect: 'The doggie is out.' But more plausibly, I think, she was referring to the dog's previous naughtiness and expulsion. Otherwise why the excitement?

A similar linguistic event took place a fortnight later, and this time reference to the past seems unambiguous. Beth went off to visit a farm with her mother's friend Amanda and Amanda's daughter, Becca. As soon as Beth got home she began to tell her mother what had happened.

BETH: Eow, tch-tch-tch. [*'Eow' means 'cat'; 'tch-tch-tch' means 'naughty'.*]
MOTHER: Did you see a cat?
BETH [*nodding*]: Eow. [*Then, pointing to her arm.*] Tch-tch-tch.
MOTHER: Was it a naughty cat?
BETH [*nodding, pointing to her arm and speaking with emphasis*]: Eow, tch-tch-tch, eow.
MOTHER: You mean that naughty cat scratched your arm? Oh dear, let's see.

A kitten had indeed made a small scratch on Beth's arm. Over the next five days, Beth told her father, her grandparents, Amanda, Becca and a man who happened to stand next to her in a carpet shop.[6]

On the basis of a systematic study Susan Foster also reports that children of about Beth's age sometimes use limited language to talk about past events. For example Kate, when she was twenty-two months old, saw a picture of a cat in a book and promptly pointed to a door leading out to the garden. Her mother said: 'Yes, there's a cat outside, isn't there, sometimes?', to which Kate replied: 'Gone.' And her mother said: 'It's gone now, yes. We saw it yesterday, didn't we?'[7]

This episode depends heavily on the mother's role, as is often the case when children are as young as this. However, it does seem likely that Kate's one verbal contribution to the exchange – the single word 'gone' – was already language in the line mode. Kate appears to be thinking back to the day before.

Foster gives one or two other examples in which the child's speech is more advanced. For instance, Ross at age two years six months is able to speak about having run around in the dark. However, the incidence of line-mode talk remains very low up to age two years six months, at which point Foster's study ended.

Other researches have shown that it remains low for a number of years thereafter, but that it increases with age and with general linguistic skill as might be expected, until by the age of five or six elaborate personal anecdotes and projections into the future have become common.[8]

The kind of speech we have been discussing arises when the child has something to communicate that she feels to be important. There is spontaneity about the utterance – and often a kind of urgency to get it out which a listener readily senses. This urgency may sometimes cause children to stumble over their words as they stretch their linguistic resources to the limit. But in general the strong desire to tell someone something has a stimulating and sustaining effect on language production. By contrast success is often more limited when the thoughts and emotions which generate such a desire are gone.

Thus Dan Slobin and Charles Welsh report that a child (Slobin's son) who said spontaneously one day: 'If you finish your eggs all up, Daddy, you can have your coffee' could only manage: 'You can have coffee, Daddy, after' when asked to imitate the first sentence ten minutes later. And work by Lois Bloom yields the same kind of evidence. For instance, in her research a boy of thirty-two months called Peter was asked to imitate: 'I'm trying to get this cow in here' – a sentence which he had spoken the day before while playing. Yet all he could produce by way of imitation was: 'Cow in here.'[9]

At first sight it may seem strange that it should be harder to imitate a sentence than to generate it. The provision of the model might be expected to help. Why does it not do so?

Slobin and Welsh suggest that, when speech is spontaneous, the 'intention-to-say-so-and-so' sustains the act. On the other hand, when imitation is called for the intention is absent and without it the words cannot be put together in the same way.

Slobin and Welsh do not consider why an intention should sustain an act. The answer seems to be that when an act is embedded in a matrix of ongoing thoughts and emotions there is a *built-in* guiding representation of a goal to be achieved – in the case of speech, a meaning to be conveyed. Without this, the child must deliberately construct the guiding representation, which turns out to be a much more difficult enterprise.

Nicholas Bernstein has argued that all action – not just human action – which is directed towards a goal must entail a 'model of future requirements (somehow coded in the brain)'.[10] In embedded action, this model is formed in a manner *experienced as effortless*. The same is true of embedded speech. No matter what effort may follow before the goal is reached, the goal itself simply seems to emerge. It does not have to be consciously and deliberately constructed.

One might suppose that an external model supplied by an adult, as when the imitation of speech is requested, would remove the need for deliberate construction by the child but this is not the case. It is of the greatest importance to recognize that the work of construction of an inner model has still to be done. (Spontaneous imitation is, of course, another matter entirely.)

The previous chapter contained some discussion of the capacity for deliberate control that develops as the intellectual construct mode begins to function. One aspect of this new competence manifests itself in language – both in its production and in its comprehension; for there arises a way of handling language, in utterance and in interpretation alike, that contrasts sharply with the spontaneous, unreflective practices considered so far. And a major difference is this: in the new, more deliberate way of functioning there is a shift of emphasis from the intended meaning to the meaning actually expressed. *What precisely is said* receives greater prominence. That is, the words acquire a new independence from context: they have to carry the meaning themselves to a much greater degree than before.

It is evident that the possession of a writing system encourages this way of using language. Words that are on a page have to be capable of standing alone. They must do without the support of gesture and intonation, for instance, or of any other non-linguistic cues.[11]

The illustration of text provides one exception to this. And in children's books pictures have an important role, for they provide some substitute for the kind of context to which the children have been accustomed since first they began to interpret speech. Interestingly when there occurs a discrepancy between words and picture, as can easily happen if artists are not scrupulously careful to respect the details of the text, then young children tend to assume that it is the text that is wrong. That is, the picture stands in for reality. But quite soon, as literacy develops, the relationship is reversed. The words come to have primacy, the pictures merely *illustrate* them. From then on, whenever there is a lack of fit, it is the illustration that is judged to be in error.

So major shifts occur in the way language is handled and regarded. Those who proceed far in our educational system come in the end to the idea of language as a formal system, with words that have defined meanings and with rules for putting words together in certain ways. And, of course, this idea is not a mistaken one. Human languages can be so considered. We can write dictionaries and grammars. We can take a sentence in isolation and ask: 'What does it mean?' But this is not what we did when we first learned to speak; nor is it the way in which even the most intellectually sophisticated adults use language in ordinary conversation. For the most part, however, when highly literate people drop back into earlier habitual ways they do not notice that they have done so.

It is important at this point not to forget the imaginative activity of context-building in which we engage when we make sense even of isolated sentences (see chapter 6). However, there remains a crucial distinction to be drawn between a manner of handling language that involves close reflective scrutiny of the *words* and one that gives priority to what the speaker intends to convey and to this end freely supplements language by non-linguistic means.

In looking at language we have concentrated so far on the latter. As to the former, prototypical instances are provided by legal language and by the setting and answering of examination questions at the higher stages of education. A student who does not 'answer the question' is in trouble. So too is an examiner who can be shown to have failed to write a question in a clear and unambiguous way.

The way of handling language that aims for unambiguous stable meaning without reliance on non-linguistic context is particularly suited to the intellectual modes because its strict use entails the separation of meaning from distorting kinds of emotion. Giving primacy to what is actually said is incompatible with reading into the words what one would like to find.

There are, of course, many kinds of language which are not intellectual in this way, yet not unplanned and spontaneous either. There is poetry, for instance. In poetry close attention to the words is essential. Each word carries weight. Yet if the poetry is of any quality, these words afford a rich interplay of thought and emotion, with many resonances of meaning.

Again, consider political speeches – or advertising slogans. Here the influencing of reason by emotion is commonly not an inadvertence but the very aim of the enterprise.

It is therefore an oversimplification to think of language as dividing neatly into two kinds. However, at this stage in the present argument, when we are trying to understand how children come to function competently in the intellectual modes, interest centres on the fact that there is a way of dealing with language which is peculiarly appropriate to that development. Previously I called this way disembedded – and so, indeed, it is. But from now on I think that the purposes of clear communication will best be served if I call it *intellectual*. For there are other kinds of language which are, to varying extents and in varying ways, disembedded too.

There are two further topics concerning the development of language that must next be considered: first, the relation of the onset of a mode to the onset of language use in that mode; and second the relation of language production to language comprehension. The two are not unconnected.

In both the point and line modes the onset of the mode takes place, as we have already seen, some time before language in that mode starts to appear. The point mode establishes itself very effectively during the first nine months of life. A nine-month-old baby gives every sign of intense feelings (interest, excitement, fury, and so on) about ongoing events; of having goals in regard to them; of thinking and learning about them. Yet language still lies ahead. Again, the line mode probably has its origins towards the end of the first year and becomes established as the second year progresses. Yet speech about the personal past and the personal future does not commonly appear until the beginning of the third year, and it remains relatively rare for some time thereafter.

So language lags behind at these early stages; but the fact is not surprising, given the magnitude of the task involved in learning about a system of such complexity. However, by the time of the onset of the third mode, between the ages of three and four, the learning has been so successful in the great majority of cases that considerable linguistic knowledge is available. So is there still a lag? Or can language from now on keep pace?

The question cannot be answered until we have discussed the relation of production to comprehension. However, in doing this a fact of some significance must be kept in mind: these two aspects of linguistic skill have tended to be studied in different ways. In general, production has been studied by the collection and analysis of samples of spontaneous speech; whereas comprehension, whenever the subjects have been old enough to make the technique workable, has been studied by the use of contrived tasks. This means that the language studied in research on production has often been in the point or line modes; whereas comprehension has mainly been studied in the intellectual construct mode. That paradoxes have appeared when results are compared is then hardly surprising.

Overall one might expect comprehension to be somewhat ahead of production on the grounds that you cannot produce what you cannot understand. And, indeed, this pattern appears when one compares studies of production and comprehension that are both in the point mode. Babies usually show signs of beginning to understand speech embedded in a supportive present-moment context several months before they start to utter words themselves. However, study after study has found that older children who can use certain words and constructions quite freely and without error in their spontaneous speech start to make mistakes when given tasks that *require* the understanding of these self-same linguistic features. The children then seem to 'misinterpret' in puzzling ways. Production appears to be in advance of comprehension. But how is this possible? How can you use successfully what you do not understand?

Allayne Bridges expresses this puzzlement when she asks: 'Could it really be that children might subsequently misconstrue an utterance which they themselves had produced *in non-test circumstances* earlier the same day?' – and answers her own question: 'Apparently so.' She reflects that this fact seems to cast doubt on the assumption, for many years unquestioned, that all linguistic processes depend on a common knowledge base.[12]

For a while I was myself driven to wonder if this reasonable assumption would have to be abandoned. I now think that we have not reached this point. The first step towards resolving the paradox which arises from evidence of production in advance of comprehension is to see the relevance of the distinctions between the modes. That is, if comparison of production and comprehension is to be valid, it must be based on data obtained within a single mode. Otherwise the differences between modes will confound it. However, even when this is acknowledged, language in the intellectual construct mode is still a complex topic.

Let me recall the main characteristics of this mode. Thought in it is dispassionate, impersonal; and it can be turned 'at will' to the topic chosen. However, there is still a need for a supportive context: thought still has to be *about* something that can be imagined. And, in general, the closer the topic comes to something directly and frequently experienced – something familiar – the easier is the imaginative task of constructing the context. This has to be moderated, however, by the evidence considered on p. 98, where we saw that an invitation to construct a quite fictitious or counterfactual context seems to be able to set the imagination free. We have to conclude that it makes a difference whether or not reality constraints are taken to be operating.

Where, then, does language fit in? What are the distinguishing marks of language in this mode?

Consider comprehension first. Here the main point is that, if language is to be interpreted in a true intellectual manner, it must achieve independence from the non-linguistic context in which it is uttered. It must be considered in detachment from the speaker's gestures and intonations – indeed, the interpretation must not depend on extra-linguistic cues of any kind. Nor must it be influenced by the feelings or biases of the interpreter, who must be capable of taking a cool look at what is actually said. That is, she must pay careful attention to the words and syntax of the utterance, laying aside any concern with what the producer of the utterance might have wanted to say or what she (the interpreter) might have expected – or wanted – to hear. On the other hand, having once attended to the words themselves in this kind of disembedded way, an interpreter functioning in the intellectual construct mode will use the words to build up an imagined context that will cause them to 'make sense' and will then use this setting in the further- ance of whatever thinking or reasoning may be called for.

When intellectual construct-mode language is produced rather than interpreted, the basic criterion of independence from the immediate context again applies. Language in this mode is relatively deliberate and planned. Also it is not an account of something that has actually happened to the speaker, nor a matter of personal anecdote. To be in the intellectual construct mode, language must share in the properties of impersonality and general- ity that characterize intellectual thought. Certain kinds of formal speech or lecture provide good examples.

Clearly children do not often engage in the production of this kind of language. They would be quite unlikely to do it on their own initiative in the pre-school or early school years; and they are not often asked to speak 'to order' on an impersonal theme during this period either, though they may be asked to give an account of a personal experience, to make up a story or to retell one that they already know.[13]

However, recent research by French and Nelson is of much interest in this connection. Their technique was to get children to talk about familiar event sequences such as getting dressed. This meant that language about non-present events was being produced on request. And although the talk was about things the children had done, it was about *general* patterns of behaviour, not about specific happenings. It can therefore be regarded as satisfying some of the main criteria for language in the intellectual construct mode. And yet the context to be constructed in imagination was very familiar indeed, so that children who were only on the threshold of that mode could reasonably be asked to undertake the task. The children who took part in the research were aged between two years eleven months and five years six months. There were forty-three of them.[14]

A further feature that makes the French and Nelson results valuable is that they concentrated on the children's use of relational terms, temporal, causal or logical – words like *before* and *after*, *because* and *so*, *if*, *or* and *but*. And some at least of these words have been much studied before in comprehension tasks of the kind that belong in the intellectual construct mode. So direct comparison seems for once to be legitimate. However, when the comparison is undertaken we discover that production is still far ahead of comprehension. The children in the French and Nelson study proved to be able to produce relational terms appropriately from about the age of four years. Yet, typically, the comprehension studies reveal difficulties and confusions that do not disappear until about age eight. Although the actual number of utterances of the critical words by four-year-olds is quite small in the French and Nelson data, it is striking that errors of the type found in comprehension studies are completely absent. For example, there are no cases where *because* is confused with *so*, or *before* with *after*.

So we have a production task and a comprehension task in the same mode and still the paradox refuses to vanish! Are we thrown back, after all, on a rejection of the assumption of a single knowledge base? Do we have to postulate one language system for production and a separate one for comprehension? I still think not. In order to resolve the paradox we must first keep the distinction between modes in mind. But this is not enough. We must next recognize that, within modes, and particularly by the time we reach the construct mode, various factors contribute to the ease or difficulty of a task.

For example, consider the findings of research aiming to assess the comprehension of *because*. One study by Morag Donaldson found that five-year-old children usually seemed to understand the word well. Yet previous research, by Emerson for instance, had reported that success did not come until about three years later; and this is a very big discrepancy. Donaldson herself suggests it may be due to the differing demands of the tasks used. She showed the children videotapes or drawings of causally linked event sequences such as a boy throwing a ball at a window, and then broken glass on the ground. Only after this did she ask for completion of the sentence: 'The window broke because. . .'

Previous studies, on the other hand, had typically just presented sentences for completion on their own. It seems very likely that the addition of visually given information greatly helped the children to construct the embedding context needed for making sense. We have seen before that this undertaking may be quite simple or very demanding as conditions vary.[15]

This last variable is relevant whether the language to be interpreted is spoken or written. But there remains a further consideration that applies when it is comprehension of speech that is at issue, as is, of course, nearly always the case when experimental subjects are under the ages of seven or eight. In such a circumstance the experimenter, being interested in *language*, generally intends that the words shall be interpreted in the typically intellectual manner. That is, attention is to be on what is actually said. However, the children are used to interpreting language in other ways. After all they learned it in the first place with full attention to all the extra-linguistic cues available. And even if now, at later ages, they *could* exclude these cues from consideration, how are they to know that they are supposed to?

The fact is that most of them discover – but only gradually – after they go to school that there are certain kinds of situation in which they are supposed to attend particularly to what words mean rather than to what people mean. They discover that the teacher sometimes, though certainly not always, expects them to do this, and they begin to learn how to comply.

Some parents encourage close attention to words before their children go to school. They talk to the children *about* words, not just *with* words. They play language games. The children of such parents go to school with a great advantage: when it comes to the interpretation of speech they are already able to move from mode to mode as the occasion demands. Their teachers usually decide that they are 'intelligent'.[16]

Most of the evidence about developments in the intellectual construct mode from age three upwards comes from research in which children are asked to tackle problems specially constructed – problems, that is, which have been devised with the aim of throwing light on some aspect of understanding. Such problems are often stated in words, even if there is also 'material' to be observed or manipulated. So the problem-solver must first make sense of the language. If this is done in an inappropriate way, then the question that the thinker tries to answer may not correspond to the one that the questioner meant to ask. It is ironic that, in such a case, a child who tries to figure out what the questioner means instead of concentrating on what the words mean is in fact, quite failing to grasp what, in a broader sense, she is *meant* to do.

When we, as adult questioners, use language in ordinary conversational ways, we expect that children will do likewise. We would be quite annoyed if they did not. But when we set someone an intellectual problem (and even a very simple construct-mode problem may be genuinely 'intellectual') we switch, often without noticing it, into using language in an intellectual way – that is, with the expectation that the precise wording is to be definitive. We intend that the language shall be given primacy over other cues as to the nature of the task. However, the trouble is that young children do not necessarily know this. It is not made explicit. Whether a given child would understand if explanation were attempted is another matter; and this other matter is evidently of critical importance in regard to the question of how the onset of the intellectual construct mode relates to the onset of language in that mode – a question that can now be seen to be very complex.

We must recognize, of course, that the question cannot even be asked unless the modes are first distinguished. Thus it has not been asked until now. The same was true of the point and line modes, but in their cases evidence already available seemed to yield an answer. This does not appear to be so when we come to the construct mode and particularly to its intellectual subdivision. I conclude that here a satisfactory answer must wait upon research of a kind that has not yet been done.

It would be wrong to suppose that failure of communication

always arises when children are asked to tackle intellectual problems; for sometimes there just is no probable interpretation of what the speaker means that clashes with the 'standard meaning' of the words. When Martin Hughes asks about numbers of children in shops or numbers of bricks in boxes then what his words mean, considered on their own, is not in conflict with any meaning likely to be arrived at when cues from the immediate context are also taken into account.[17]

However, this is not always so. And sometimes it is even the case, when psychologists set out to study the development of thinking, that they contrive situations where the immediate context exerts a strong pull on the interpretation of the questions asked. For instance, this is true of much of the research used by Piaget to establish the distinction between 'pre-operational' and 'concrete operational' thinking. The child generally faces a kind of temptation to answer wrongly.

Students, when they first learn of this body of work, are apt to grow indignant. 'It's not fair,' they say. 'The child is being tricked.' However, the complaint is not legitimate. There is nothing improper about research aimed at discovering when, and how, children start to resist intellectual temptation and answer in a new way. On the other hand, it is crucial to be clear about the nature of the temptation and about what is entailed in overcoming it.

Consider an example. Piaget shows children a bunch of flowers in two colours: say, five red and three yellow. He then asks: 'Are there more flowers here or more red flowers?' Children below the age of six or seven tend to answer: 'More red flowers.' If they are asked to justify this response, they most commonly say: 'Because there are only three yellow ones.'[18] It is quite clear that they have based their answer on a comparison of the red flowers with the yellow flowers. Yet they were asked to compare the red with the whole bunch. Why did they not do as requested?

Piaget thinks the reason is that they could not, and this because of limitations in their ability to handle the relations between a class and its subclasses. Specifically, he holds that a child of less than seven or thereabouts cannot think of a whole class and one of its subclasses at the same time. If the child 'centres' on the whole class, the notion of the subclass is temporarily lost, and vice versa. This, if it were the case, would clearly make comparison of class with subclass impossible. The child could not deal conceptually with the inclusion of one class in another.

It is an intriguing explanation, which has attracted a vast amount of interest. However, if one accepts this as the only source of error, one is led to predict that, when there is no *inclusion* relation between sets of things to be compared, no similar difficulties will arise. And this is not so. James McGarrigle conducted a series of studies of which the following is an example.[19]

The children were shown sets of toy cows and horses, black or white in colour, arranged thus, on two sides of a toy wall:

	Cows		
B	B	W	W
B	B	B	W
	Horses		

They were then asked such questions as: 'Are there more cows or more black horses?' – a question having the same syntactic form as: 'Are there more flowers or more red flowers?' but referring to a very different state of affairs. For now there is no possible overlap between the two classes to be compared. Yet still many children make the wrong comparison. Commonly they compare the black horses with *black* cows, saying things like: 'There's more black horses 'cos there's only two black cows.'

Thus McGarrigle's work reveals a kind of misunderstanding which has nothing to do with handling class inclusion but which has obvious relevance to Piaget's class-inclusion tasks. I must emphasize strongly, since I have been misunderstood before, that I am not claiming this to be the *only* source of error. What I am saying is that it is an important source of error of which Piagetian theory does not take account. What the error amounts to is a failure to hold to the precise wording of the question asked. Instead of doing this, the children, to borrow the words of

Inhelder, Sinclair and Bovet, substitute a 'more natural' question of their own.[20]

It is far from easy to say what makes one question 'more natural' than another, but there is no doubt that, in a given context, some questions do seem more natural than others to human beings, young or old.[21] This is why students, when they learn about the standard class-inclusion question, are apt to think of it as a trick. Even adults are at first inclined to answer: 'More red flowers.' But if they are asked to listen again, they say: 'Oh! I see. More *flowers*, of course'; whereas young children normally persist in their first interpretation. This is because the natural question arises from the words in context, not from the words considered alone.

Much evidence is now available to show how shifts in the immediate context can occasion shifts in language interpretation. As an example, consider the findings of some research carried out by McGarrigle and myself.[22] The children who took part in this study were shown two rows of toy cars, one containing four cars, one containing five. The rows were arranged on two shelves, one above the other, to make comparison easier. The extra car always projected to the right in a noticeable way.

The children were asked: 'Are there more cars on this shelf or more cars on this shelf?' And usually the question was answered correctly. Then a change was made to the arrays. With the children watching, a set of enclosing garages was placed over each row. (These had no floors and could easily be put in position without disturbing the cars.) The four-car row was given four garages; the five-car row was given six garages. So now on one shelf the garages were full, while on the other an empty garage was plainly visible.[23] The question was then repeated.

Now the main thing to notice is that the change introduced was not relevant to the meaning of the question. The addition of the garages in no way affected the numbers of the cars. If there were more in row x to begin with, then were more in row x still. However, the children did not always seem to think so. About a third of them now told us that there were more in the four-car row – that is, there were more in the row that was full. The immediate context was exerting some kind of pull on the interpretation of the language.[24]

Shifts in the interpretation of language as the context shifts are not at all rare. In everyday speech they are indeed normal – and not just for children. Adults are sensitive to context too. The question: 'Have you put all the knives and forks on the table?' will be varyingly interpreted according to whether the purpose is to tidy out a drawer or to lay the table for a meal. The meaning given to 'all' will shift, as did the meaning which the children gave to 'more'. The research findings seem surprising only because the effect is usually not noticed.

To take another example, consider the sentence: 'The gun went off at ten o'clock, so we had a good night's sleep.' This will seem odd to a reader treating it out of context, except to one who has had the kind of personal experience that will allow an appropriate act of imaginative embedding quickly to take place. In some country districts farmers use gas guns to scare birds away. The guns 'go off' regularly, at short intervals, and unless, in another sense, they 'go off' as darkness falls, people in the neighbourhood spend a disturbed night. With the appropriate kind of embedding, the words: 'The gun went off at ten o'clock' will be taken without hesitation to mean that the gun stopped firing at that hour, though this is a relatively uncommon interpretation and indeed is directly opposed to the common meaning.

Spoken language, then, is ordinarily used in a way powerfully influenced by context, and this is entirely reasonable and sensible for purposes of everyday 'face-to-face' interaction between human beings. However, as we have seen, there are other human purposes for which intellectual language is essential: legal purposes, for instance, and certainly the purposes of strict reasoning and science. In these activities words must have agreed and stable meanings as far as possible. They must have precise definitions that are respected by users and interpreters. Shifts in meaning, if they occur, must be made explicit: they must be announced. This at least is the ideal, though it is often not attained.[25]

In the history of the human race the advent of language that meets these intellectual needs is almost certainly consequent on the invention of writing. Language written down has a physical permanence which gives it independence from the fluctuations of context and which makes it more available than spoken language for reflective analysis.

The same relationship surely obtains in the life of a single human being. The process of learning to read undoubtedly encourages the decontextualized treatment of language, especially if reading is taught in ways that encourage close attention to precisely what is there on the page.[26]

Some facility in the handling of intellectual language is a precondition for the later manifestations of the intellectual construct mode. These developments are extremely wide-ranging. They have, for instance, a crucial part to play in science and inventive engineering.

Before considering them, however, it is necessary to discuss the next, and final, mode in the intellectual line of development.

8

The Intellectual Transcendent Mode

We move on now from the intellectual construct mode to a further mode that is also intellectual, yet radically different. I shall call it the intellectual transcendent mode. These two modes are alike in that thought has primacy. I stress again that this does not mean they are unemotional. It means only that certain kinds of emotion – the kinds liable to distort or bias thinking – are excluded by definition. For where potentially distracting emotions are present to any appreciable degree the mind has shifted gear and has slid back into one of the core modes – a shift that remains always easy and tempting.

However, the kind of emotion that arises from a sense of the importance of achieving some new measure of understanding is entirely appropriate to the intellectual modes and can be exceedingly strong. Indeed, it is normally so.

The difference between the two intellectual modes lies in the locus of concern. In the case of the construct mode the locus of concern is still within space-time, though not restricted to specific episodes therein (as it would be in the point or line modes). In the intellectual transcendent mode, however, the concern is no longer about something that could happen 'sometime, somewhere'; and the need for an imagined setting has dropped away. The mind is now finally able to function without recourse to contexts derived, either directly or by imaginative extension, from personal experience of living in the world.

The last sentence is, I think, liable to give rise to misunderstanding unless I stress what I do not mean by it. I am not saying that personal experience of living in the world is irrelevant to the advent of the intellectual transcendent mode. Indeed, I would consider such a claim absurd. What I am saying is that certain kinds of contextual support are no longer needed. The mind has become able to function without them.

This is the final stage in the long path from the extreme of embedding at one end to the extreme of disembedding at the other. Along this way each new mode achieves some new measure of disembedding, some reduction of contextual binding by comparison with its predecessor. For example, the construct mode, in whatever variety, is less contextually bound than the line mode since the locus of concern can be 'anywhere'. Yet the construct mode still needs links with the known. With the coming of the fourth mode the last requirement is dropped. So, if we take up again the metaphor used for naming the first two modes, we may say that the fourth mode is 'spaceless'. It needs no local habitation, present, remembered, foreseen or imagined. To speak paradoxically, we may say that the locus of concern is nowhere.

To say this, however, is very far from saying that concern is with nothing. It is necessary now to give a positive account of the kind of mental functioning that the new mode brings.

The prototypical activities of the intellectual transcendent mode are logic and mathematics. But since we are concerned with concern, the question we have to ask is what, in the most general sense, are logic and mathematics *about*?

Such answers as *number, shapes, syllogisms, propositions* or the like will not do; for they are too limited and particular. The general answer has to be that logic and mathematics are about *relationships*: relationships of compatibility or incompatibility, of symmetry or asymmetry, of inclusion or exclusion, of equality or inequality, and so on. More than this: they entail the systematic study of *patterns* of relationship. And what we call 'creative mathematics' is the attempt to extend this study so that its previous limits are surpassed. That is, a known pattern is extended or new patterns are revealed.

At this point the objection may be raised that 'relationships' occur in space-time, just as much as 'things' or 'happenings'. So in a sense, they do. But when concern shifts from 'things-in-relation' (or 'events-in-relation') to the relations themselves this is a major shift, with major preconditions and major consequences. One critically important consequence is that the patterns, as distinct from their embodiments, are not bounded by the limits that space-time imposes. They can be extended into n dimensions, to infinity or to eternity.

There are certain relations that have multiple instantiations in space-time. Some of these instantiations are familiar to all human beings from an early age. For instance, the bony structure of our own bodies is (roughly) symmetrical about a vertical plane. But it is one thing to be interested in one's own body, or even bodies in general, and another thing to be interested in symmetry.

Again, we all experience early in life the relation of incompatibility. For example, this can take the form of a clash between what we ourselves desire and what someone else desires. Such clashes are of great significance to us. We all notice them and care about them. But it is one thing to recognize this kind of incompatibility in action, even to the extent of thinking about it and understanding how it arises (which entails having an awareness of other independent minds). It is a different matter entirely for concern to focus on the relation of incompatibility itself.

A shift of concern from things-in-relation to relations themselves is made easier if the dominance of the 'things' can be diminished. But that dominance is powerful indeed, and hard for our minds to reduce. It is harder still to give up entirely the construct-mode habit of thinking about imaginable entities of some kind, even if their individuality is diminished to the limit. This means that the diminishing of individuality is not a sufficient condition for the shift to concern about relations, though I think it is a necessary one.

The obvious way to think about things while reducing their individuality is to use some means of schematic representation. It turns out that schematic representation presents no great problem for children of three and over. Indeed, it is resorted to quite readily by some of them when dealing, for example, with problems of a numerical kind. Hughes reports a study in which children aged between three and seven were given paper and pencil and were shown a number of bricks – varying between one and six – lying on a table. The children were asked to 'put something on the paper' to show how many bricks there were. Of the three- and four-year-olds, about one-third used an arbitrary mark to stand for each brick, and the mark chosen was often a simple line or tally.[1]

It should be added that the frequency of arbitrary marks fell off at age five, at which point there was, unsurprisingly, a sharp increase in the use of numerals. But there was also a sharp increase in straightforward drawings of the blocks, which looks like a movement away from any attempt to diminish the salience of the things themselves with their physical properties. Perhaps the early schematic representations were used mainly because actually to draw the blocks seemed like a hard undertaking. It remains true to say that these young children showed themselves able to think of using tallies.

Evidence of the use of schematic strategies was also found among older children. Here is one instance. The problem was: 'Twelve children are going home on the school bus. Five children get off. How many are left on the bus?' Kim, who was six years old, answered correctly. She then said that, to work it out, she 'thinked' in her head; and she explained that the thinking involved using 'dots' in her head.

In the study that yielded this example Hughes was still interested in written representation, but Kim did not spontaneously put her dots down on paper. Indeed, she later showed some resistance to putting anything on paper at all, though she was able – with varying success – to use the conventional numerical notation that children learn for 'doing sums'. In this she was something of an exception. Hughes's work led to the general conclusion that the formal code of arithmetic ($12 - 7 = 5$, and the like), as learned during the primary-school years, is very apt to become 'a system of rules and procedures divorced from concrete reality'.

Now if this meant that concern had moved from 'things' to

'relations', then it would be entirely appropriate as evidence that the foundations of mathematical competence were being laid. But there is little reason to interpret the findings in this optimistic way. If you understand a pattern of relationships that has instantiations *of a highly familiar kind*, you will be able to see how the pattern 'works', or is applied 'concretely'. But many children, even if they could 'do sums' correctly, seemed not to see the bearing of the arithmetical expressions on relations between sets of ordinary everyday things. For example, to the problem: '7 − 4 = ?', Nadine gave the answer: '3'. She was invited to 'check' this, using bricks. She built a tower of seven bricks, laid out a row of four and said 'take away'. She then removed the row of four and wrote: '7 − 4 = 7', appearing to be quite happy with the outcome.

It is possible to argue that the invitation to check the answer led Nadine to think that it was wrong. But no child who had a reasonable grasp of what she was being asked to do could possibly have responded in this way. The conclusion seems inescapable: she did not understand the 'minus' symbol; or she did not understand the 'equals' symbol; or she understood neither.

Hughes reviews evidence about the 'equals' symbol, evidence which shows that children treat it more as an instruction to 'do something' – add, subtract or whatever – than as an expression of equality. For example, Behr and his colleagues found that children tended to regard statements like: '3 + 2 = 2 + 3' as not legitimate. One boy suggested amending this one to read: '3 + 2 + 2 + 3 = 10'.[2]

In interpreting these findings, Hughes stresses the difficulties of learning the mathematical *code*. He treats the children's confusion as evidence of failure to 'translate' – in either direction – between this code and representations of 'concrete reality'. And I think he shows clearly that large numbers of primary-school children have indeed not learned to make this movement freely. They may know how to carry out the formal procedures with some success, though with many failures too; but in either event they often show little understanding of what they are doing.

The general solution which Hughes proposes is to help the

children to use appropriate images more effectively.[3] This amounts
to saying that, for some time after the teaching of the formal code
begins, the children should continue to function in the construct
mode. They should not be expected to understand the relations
expressed by that code without the support of illustrative embodiments which make sense to them.

I believe this to be wise. However, it is necessary not to lose
sight of the further goal: the shift from the study of things-in-
relation to patterns of relations themselves, which is the shift into
the transcendent mode.

Hughes observes that many eminent mathematicians have found
images helpful. Bruner and Kenney make the same point: research
mathematicians sometimes report that they use 'concrete props' in
their thinking. But we must remember that this is not the whole
story.[4]

Bruner and Kenney studied 'in minute detail' a small group of
intelligent eight-year-olds who were being taught mathematics
intensively by a research mathematician, aided by 'tutor observers',
one for each child. The children were encouraged to use concrete
– that is, physically present and manipulable – materials to help
them; but they were also encouraged to develop or adopt suitable
notations, and then to look for other fresh embodiments of the
same mathematical idea. In these favourable circumstances,
'continous interplay' occurred between formal code and physical
instantiation. But Bruner and Kenney note that 'the problem now
is to detach the notation . . . from the concrete, visible, manipulable
embodiment to which it refers. . .'.

And why must this further step be taken? They go on to state
succinctly the crucial reason:

For if the child is to deal with mathematical properties he will have to
deal with symbols *per se*, or else he will be limited to the narrow and
rather trivial range of symbolism that can be given direct (and only
partial) visual embodiment. Concepts such as x^2 and x^3 may be given a
visualizable referent, but what of x^n?

Hughes is absolutely right about the need to avoid what he calls the 'dangerous gap' between the formal code and the known physical world. And he is talking realistically about what might be done in schools to reduce it. Bruner and Kenney emphasize that in their study both the children and the learning situation were far from typical. However, the general point that they make remains valid: there are limits to the possible embodiments of patterns of relationship. Or, to put it differently, the human mind can – in principle, though not easily – conceive of systems of relations which extend beyond anything known or imaginable in the world of space and time.

This remarkable fact holds even in Euclidean geometry where it might seem least likely to be true. For example, in Euclid's *Elements*, one of the definitions is: 'A line is breadthless length.' But we never encounter and cannot actually imagine breadthless length. Whitehead, on the same general theme, says, with the authority of his eminence as a mathematician and a philosopher that 'for example, the shape-iness of shapes, e.g. circularity and sphericity and cubicality *as in actual experience* [my italics], do not enter into the geometrical reasoning'.[5]

In spite of the nature of some Euclidean definitions, this view of mathematics, though now widely accepted, is fairly recent – that is to say, it is largely a product of the nineteenth and twentieth centuries. Kneebone gives a clear and interesting account of how thinking on the subject has developed, arguing that geometries, whether Euclidean or non-Euclidean, are to be thought of as abstract theories. Whether any one of them applies to space as we know it then becomes an empirical question to be settled by the ordinary methods of scientific inquiry.[6]

Now the intellectual transcendent mode is certainly concerned with what are, in this sense, abstractions. But, from the point of view of developmental psychology, such a way of putting it is not very helpful. We want to know – among other things – what changes take place when this mode becomes available to the mind. 'Abstraction', however, is far from being a new activity. To return to the example used earlier, a child of eighteen months – or younger – who can point to a blob of dark wool on the face of a teddy bear and say 'nose' is dealing in abstractions – and abstract relationships at that, for the 'nose-ness' of the blob depends on its position in relation to the other features. Exactly the same blob detached from these features would not be taken for a nose at all. However, the young child who demonstrates this skill with abstraction is interested in the nose, not in the relationships. Many years will pass before his mind starts to concern itself with the latter, considered apart from entities or events which he experiences or can imagine. Indeed, his mind may never develop this concern to any significant degree. Many minds – many normal, not to say lively, intelligent minds – never do.

We shall have to ask why some do and some don't. But first: at what age does the intellectual transcendent mode begin to appear? What are the earliest signs of its advent?

My observations of children – mainly in Scotland, but also in the United States – lead me to suggest, for these cultures, an average age of around nine. But let me emphasize that I am talking about first indications of this mode, not by any means about fully developed competence.

One very telling sign of change is an increase in systematicity. Consider a task devised by Piaget and Inhelder in which a child is given sets of small cardboard squares, each set having a different colour.[7] Thus there might be a pile of red squares, a pile of blue, a pile of yellow, and so on. The child is then asked to make as many pairs as possible, with the constraint that each pair is to contain two colours and no two pairs should contain the same colours. In other words, as many combinations as possible are to be produced.[8]

Children younger than seven or thereabouts tend to choose cards in an unsystematic way and then stop when no new pairings readily occur to them. Asked, at this stage, how they know they have found all the pairs that they could make, they say things like: 'I looked and I saw.' And they seem perfectly satisfied. Seven- or eight-year-olds are also often unsystematic but are more likely to show signs of unease when pressed about whether all possibilities have definitely been found. Around nine, however, some attempt to be systematic is usually made, even if it does not fully succeed; and the importance of system as a way of exploring relationships and 'being sure' about them comes to be recognized. Beyond this there remains still the crucial discovery that a systematic approach can yield a way of working out the possible combinations for any number. That is, the child can come to realize that he does not need to have the sets of cards in front of him in order to discover the combinations. He can devise a formula and then use it for any number of sets at all. I have quite often seen children of ten or eleven achieve this, before they have been taught anything of the kind in school. When this happens, they usually also show signs of being very intrigued by their own discovery. And no wonder.⁹

The recognition that a pattern of relations can be extended – sometimes indefinitely – is of the essence of transcendent-mode thought. And delight in this kind of realization is of the essence of the attendant emotion. I know of one very advanced child who seemed to have such an experience at the age of seven years six months. He was asked how many parts there would be if an apple was cut in half and, having answered *two*, he then added spontaneously: 'And if you cut those in half you'd have four, and if you cut *those* in half you'd have eight, and if you cut *those* in half you'd have sixteen, and then – poof!' As he finished he waved his arms in the air and smiled with pleasure.

I have asked children on occasion if there is a number that is the biggest number possible. Many, at the age of seven or eight, think there must be, though they don't know what it is. One girl told me she reckoned it would 'fill this room'. (I failed to discover quite what she meant by this.) But around nine or ten there often comes the idea of a number series extending without limit. Children then make remarks like: 'You could always go on adding another one.' Thus some notion of infinity is born. And occasionally one may witness the birth in the form of an 'Aha!' experience as the child suddenly sees that the possibility of adding one more can never, in principle, reach an end.

In all of this kind of thinking, concern with what is possible (given certain initially accepted constraints) is evident. Piaget would say that this is characteristic of what he calls 'formal operational thought'. He considered various ways of capturing its essential attributes but came in the end to conclude that it 'implies the subordination of the real to the realm of the possible and consequently the linking of all possibilities to one another by necessary implications that encompass the real, but at the same time go beyond it'.[10]

I agree with this. It could not be better expressed. However, I do not like the name 'formal operational' because it arises from aspects of Piagetian theory that I cannot accept. His use of 'operational' is inseparable from the claim that this kind of thinking depends on the development of certain specific *structures* in the mind – structures that Piaget attempts to describe by the 'algebra of logic'. It would be inappropriate here to criticize this claim, for that could not be done adequately and fairly without lengthy discussion.[11] I shall therefore say only that I do not think the psychological reality of these 'structures' has been established, and I have not found in them a useful explanatory tool.

In order to think effectively about patterns of relationship it is necessary to manage one's mind. Thought must be directed in a controlled and disciplined way. This is the case even where the pattern to be explored is quite limited in scope. In his account of what is entailed in the 'subordination of the real to the realm of the possible' Piaget insists on the importance of being able to explore systematically the combinatorial possibilities that arise when one is dealing with two-valued propositions – that is, propositions which may be either true or false. I once studied the ability of young children to explore a very simple and limited relational pattern of this kind, presented in a familiar context of course, since the children were only about six years old.

Each child was shown a doll's house with a path leading to the door and was told that Mr and Mrs Brown lived here. The child was then invited to imagine that she and I were going up the path to pay a visit. The question was: who might be at home? The game took the form of repeated visits to the house, on each of which the child was asked to say whether we might find a new situation this time. That is, suppose that Mr and Mrs Brown had both been deemed to be in on the first occasion, could it then be different when we next called?[12]

Notice that the task involved two propositions: 'Mrs Brown is at home' and 'Mr Brown is at home', each of which could be either true of false. If we call these propositions 'p' and 'q', then the possibilities were: both p and q true; p true and q false; p false and q true; both p and q false. No dolls were provided to represent the characters. Apart from the house there was no physical support for the child's constructive consideration of possibility.

It turned out that, in this context, most of the children could easily work out the four possibilities, even the one where both p and q were false – which is to say that no one was at home. What many children did not do, however, was decide then that all possibilities had been exhausted. Instead they often invented new characters: a daughter or a son, a friend who had come to stay – even a dog! Thus they revealed an attitude very common at their age: they seemed to see no need to accept the limits imposed by what was 'given'. Their thinking was not constrained in any strict way by the information that I had provided. Another way to put it is that they did not have a clear conception of *this problem* – this one and no other – which they could hold on to and use in deciding when the problem had been successfully dealt with, so that thinking about it should cease.

Such a conception is the very foundation of relevance. And there can be no intellectual power where a sense of relevance is lacking. This applies equally to a problem which one sets for oneself and to a problem which one accepts already formulated, from another person.

Without the rigour and control entailed in respecting all – and only – the conditions of the problem (the conditions, that is, which make it *this* problem and not another one), then the activities of the intellectual transcendent mode cannot be pursued effectively. The fun and freedom involved in inventing a new occupant for the Brown household have to be given up. There arises, however, a new source of fun and freedom in place of the old. (Also, of course, the old kind can be given full scope in other ways, since new modes do not cause older ones to be abandoned.)

The question now to be considered is the nature of the link between precise delimitation of 'the problem' (or topic for thought) and the movement from thinking about things-in-relation to thinking about the relations themselves. I am sure that these are closely connected but I am not sure how. It might be that the ability to demarcate a problem – and hold on to it thereafter, without letting it slide and change – is a precondition for the coming into prominence of relational structure. After all, the world as we experience it, the world of things and happenings, keeps changing; and while the mind is caught up in this flux, either directly or imaginatively, reflection on systems of relations is no doubt difficult or even impossible. But on the other hand, one might perhaps argue that it is the advent of interest in relations which brings with it recognition of the importance of demarcation, that is, of having a precisely specified set of relations to explore.[13]

Most likely, given what we know about human development in general, there is interplay. Thus as problems that belong in the intellectual transcendent mode are more clearly stated and apprehended, interest in the discovery of relational systems grows, and as interest grows the importance of holding to 'this problem and no other' comes to be more fully appreciated.

If this is so, then the interplay will lead to a gradual extension of the complexity of the relational systems that can be effectively explored. But we come now to a further consideration, namely the importance of notation. It is impossible to deal with systems of relations, beyond the very simplest kind, without some way of writing them down.

We have already taken stock of the fact that children are able, at an early age, to develop for themselves ways of representing on paper the number of objects set out before them. It turns out, somewhat surprisingly, that they do not even find it hard to devise a symbol for zero. But Hughes, who reports this finding, also reports that it was very much more difficult for the children he studied to represent transformations or relationships.[14]

He tells us that, when asked to show that certain objects had been added to or subtracted from a pile, most children did nothing to differentiate these two operations. Only eleven out of ninety-six children (aged between three years four months and seven years nine months) even attempted to mark the distinction between adding and subtracting and only four did so in a way which might have been understood by an uninformed observer. Among the remaining seven who did at least try there was, however, a considerable display of ingenuity. For example, Scott, who was one of the oldest children, indicated addition by drawing the appropriate number of British soldiers marching from left to right and subtraction by the use of Japanese soldiers marching from right to left! We may admire Scott's ingenuity; but his notation would be cumbersome in use.

However, the culture in which Scott was being brought up provides a very good notation; and Scott had been at school for more than two years being taught about the use of 'plus' and 'minus'. Why did he not use these symbols? Why did no single child think to use them? The nearest anyone came to their use was Christopher (also among the seven-year-olds) who wrote 'took 1 away' and 'add 3'. But this is still some way from using the arithmetical code.

I return to the account by Bruner and Kenney of the small group of privileged eight-year-olds. Bruner and Kenney say: 'What was so striking in the performance of the children was their initial inability to represent things to themselves in a way that transcended immediate perceptual grasp.' Later, however, as learning progressed, the children did seem to come to understand abstractions 'expressed in a common notation', while at the same time developing a good 'store of concrete images for exemplifying the abstractions'. But these children were receiving very special individual help and attention, as we have seen.

★

Much remains obscure about how the intellectual transcendent mode comes upon the scene and by what means competence in it develops. I have tried to indicate some of the conditions that seem to be necessary. To recapitulate, these are: having a way of reducing the prominence of things and increasing the prominence of relationships; having a firm sense of relevance, which means demarcating and holding to 'this problem and no other'; developing an understanding of the value of proceeding systematically, together with some skill in doing so; and having available for use, with understanding of its function, a written notation well fitted to the particular pattern of relations that is to be explored.

There may be other important conditions that I have left out. But one thing is already clear: few people, if any, would achieve much of this without a great deal of help. The help depends first of all on the existence of a cultural tradition. Beyond this, it must take the form of initiating individual children into the tradition. It is a matter of education.

PART TWO

RANGE AND BALANCE IN MATURITY

9

The Intellectual and Value-sensing Modes: A Look at History

It is time now to return to the question mentioned in chapter 1 and raised again at the end of chapter 6. In intervening chapters we have looked at the nature of the modes called 'intellectual'; and we have seen that in them emotion is by no means without its role. Passionate curiosity empowers the intellect. Also the achievement of new understanding is normally accompanied by delight. The intellectual modes are marked, nevertheless, by an experienced distinction between thought and emotion and by a measure of control that makes possible the exclusion of certain kinds of emotion, namely those that are inappropriate because they are liable to distort thought. This is a matter of definition. Where these criteria are not satisfied, the mode is not intellectual in the sense used here.

The question to which we now turn is whether human beings can achieve any analogous development of the emotions. What would such a development be like? Is there any evidence that people can function in modes which parallel the intellectual modes but in which emotion, rather than thought, has primacy?

Let me first restate a central tenet: all emotion is evoked by the apprehension of importance. Where nothing matters, there is no emotion. So if there prove to be emotional modes which are analogous to the intellectual ones, then appropriate sources of importance must be entailed. And just as thinking in the intellectual modes is not concerned with line-mode happenings – specific events in specific lives lived – so emotion in the parallel modes (if they exist) must not derive from such happenings. It must not depend on the success or failure of line-mode purposes. It will have to be evoked in other ways.

Beyond this, there will have to be the same kind of experienced distinction between emotion and thinking as in the intellectual modes, and the same kind of control. However, what is to be regulated by this control will be thinking of a kind that might interfere with emotion rather than emotion of a kind that might interfere with thought.

Now we have to consider the possibility that the development of modes which satisfy these criteria might in principle be alien to us. Emotional development might reach some necessary limit before that point, leaving us with a major asymmetry between the emotions and the intellect. It is often remarked that such an asymmetry is characteristic of modern societies, where progress in science and technology has led to a dangerous imbalance, so that we have become one-sided. And there are indeed grounds for this claim.[1] If it is well founded, we need to understand how the asymmetry has arisen and what, if anything, can be done about it.[2] What has made the intellect so powerful? Is it really true that there have been no comparable developments affecting the rest of the mind? And above all, if not, why not?

At this stage it seems necessary to look to history. We need to know what has been achieved – or tried – by human beings before us. We need this knowledge to take stock of where we stand now.

The fundamental question is whether we can find clear evidence of human emotional responsiveness to sources of importance that seem appropriate to the modes we are investigating. If the answer to this is 'no', then the inquiry is effectively finished. If the answer is 'yes', further questions will have to follow.

A very brief reflection is enough to show that evidence for the existence of the modes we are wondering about might be found in at least two places: in responses to art and music, and in certain kinds of experience that we call religious or spiritual. I shall concentrate at this point on the latter because it turns out that there are relevant distinctions which emerge there quite clearly. At the same time I do not want to imply that only religious or spiritual experiences are relevant. I shall call the modes we are discussing *value-sensing modes*, with the proviso that the values in question must transcend personal concerns.

All the world's great religions are infused with a sense of self-transcending values – values which seem to emanate from 'beyond' us and yet to call for human response. It is not relevant here to ask whether this sense is the recognition of something 'real'. What matters – for now – is just the fact of the human experience. And this has certainly been widespread and powerfully emotive.

So do we have evidence already for the existence of emotional analogues to the intellectual construct and transcendent modes? Not quite. Recall that these modes entail a large measure of separation of thought from emotion: they are *dispassionate* (though this has to be qualified, as we have seen). So genuine analogues on the emotional side would have to be, in corresponding measure, *acognitive* – that is, thought and emotion would again be separated, but this time emotion would be the primary function.

We have to recognize first of all that religion, especially in its early forms, has very often been closely combined with some kind of attempt to explain things, often amounting to a cosmology, so that thought and emotion have not been separated. Thus it has arisen as a manifestation of the core construct mode, where thought and emotion remain tightly interwoven, the goal being to understand the nature of reality *and* respond to it emotionally in an appropriate way – a goal which, on the face of it, is by no means unreasonable. So there have developed systems of explanation and forms of emotional expression in close connection.

Ancient civilizations yield very many examples of this. Generally the systems of explanation take the form of narratives of the kind we call myths. These myths commonly tell stories that account for the origin of the world and for some of its particularly striking characteristics. The characteristics that seem most striking to given people will obviously vary according to the conditions of their lives. For example, in Egypt we find myths about a creator-God called Atum who arose from the primeval waters perched upon a little hill. In Egypt every year, as Plumley points out, the flood-waters of the Nile recede and there emerge little hillocks whose tops, being very fertile, soon swarm with life.³ On the other hand, in Scandinavia altogether different themes emerge:

The myths have a sense of mighty spaces, of darkness and terror and tremendous barriers confronting travellers, human or divine, on journeys from one realm to another. There are great mountains and rushing rivers, like those of Norway and Iceland, blocking the road. The feeling of immensity, darkness and cold is in keeping with the long cruel Scandinavian winters with their brief hours of daylight and savage blizzards. Yet as if reflecting the restlessness of the Viking Age, there is constant journeying between the worlds. The road from the land of the gods to that of the dead took nine nights through deep, dark valleys, over rivers welling up from the lowest depths . . .⁴

Thus the human imagination, here as always, uses what it knows well in pursuit of efforts to move towards what it does not yet comprehend.

Some characteristics of the world are, of course, of outstanding significance to human beings no matter where they live. Thus, for instance, we find over and over again attempts to explain the observed rising and setting of the all-important sun. Not uncommonly there is some concept of a heavenly boat or chariot in which the sun travels from east to west, and it is then – reasonably enough – often supposed that the journey must continue under the earth during the night so that the reappearance can duly take place at dawn. This kind of idea is found both in Ancient Egypt and among the Scandinavians, in spite of the great differences between their cultures.

Sun myths, however, do not always entail an underground journey. One Egyptian version tells how the sun is born each dawn as the child of the sky goddess, reaches manhood at noon and sinks down in old age into the west. And there is an Indian myth according to which the sun travels back from west to east across the night sky shining upwards. That is, the sun is conceived as having a bright side and a dark side, one being turned upon the earth by day and the other by night.

It is evident that when these myths were conceived there was already a rich capacity for the generation of hypotheses; but they were hypotheses of a kind not susceptible to checking: they could neither be confirmed nor infirmed. Their role as attempts at explanation is very obvious but they were at the same time usually central to religious practices. To return, for example, to the charming image of Atum on his hillock, we hear that at Hermopolis a hill rising on an island in the middle of a pool was a place of pilgrimage and a centre of ritual.[5] It is then not surprising to find that different priestly sects made competing claims as to the site of the Primeval Hill. For the Primeval Hill was holy. It was an object of worship, calling forth emotions of awe and veneration.

Mingled in all this, traces of the functioning of the line mode are clearly to be seen – for instance, in the competing of the sects. And again there is the narrative structure: explaining the world is a matter of telling *what happened* – or what still happens, like the sun's daily journey – not of discovering 'laws of nature' as these came later to be conceived. However, there is also evidence of attempts to deal with some very difficult ideas: ideas such as timelessness, nothingness and infinity. For example, Egyptian creation myths postulate that, before the beginning of things, there existed only the 'Primordial Abyss of Waters' which was limitless and without directionality – 'in the infinite, the nothingness, the nowhere and the dark'.[6] The lack of what we might regard as intellectual sophistication was thus by no means total. We must also recognize that the extent to which the myths were taken literally probably varied a great deal, not only across cultures but within cultures. At all times some people were no doubt more ready than others to regard the stories as actually 'true'.

A further – and important – question concerns the extent to which the myth-laden religious systems were seen as serving line-mode purposes, that is, as helping to bring about specific desired events. Certainly such purposes were often part of the package. Recall that the line mode and core construct mode generally function in close association. Thus the powers postulated in the myths were to be worshipped not just because of their importance, not just with the emotions called forth as the due response to this importance, but so as to achieve a successful hunt, a good harvest, a victory over enemies.

But was this all? If the answer is 'yes', then the worship of the gods in such systems was just an elaborate, fantastic way of trying to influence the course of events in time.

It seems likely, however, that this was not all. More probably there was some genuine, but no doubt fluctuating, sense of the sacred – an inkling at least of that 'vision of something which stands beyond, behind, and within the passing flux of immediate things' which Whitehead regards as the essence of religion. He goes on: 'The immediate reaction of human nature to the religious vision is worship.'[7] And it seems unlikely that the very idea of worship should ever have arisen without some vision of this kind, however mixed with other things.

Cassirer claims that when magic arose as a component of early religion it was not typically used in everyday matters to produce practical effects. Rather it was reserved for 'bold and dangerous enterprises' – for very special undertakings.[8] We learn from Plumley that *Hike*, the early Egyptian form of religious magic, was supposed to be used primarily to protect the State and the temples; but it was also considered legitimate to use *Hike* to protect people from illness and to help the sick and the dead. On the other hand, its completely free use for personal ends was much frowned upon. Plumley reports that this was regarded with 'repugnance'. It seems to have been considered an impiety, an abuse of the holy.

We have, then, much evidence for the widespread occurrence in ancient cultures of religions that arose principally from the functioning of the core construct mode. Now we must go on to ask, in a historical context, questions about the other variants of this mode in which thought and emotion are no longer so closely interwoven. And although we are primarily interested here in developments of the value-sensing modes we are also concerned with the modern disparity and whether it has always existed. So it may be a good plan to take a brief look first at what can be said about the origins of the intellectual side of the division.

At once it has to be acknowledged that evidence about absolute beginnings is impossible to find. We have seen that, nowadays, children from about the age of three can tackle problems which, though simple, are genuinely in the intellectual construct mode. I am not committed to the view that individual development always follows the course of development in the species. Nevertheless, the ease with which children take to this kind of activity today suggests that it was probably not inaccessible to our quite remote ancestors. And, of course, as usual, different modes no doubt existed side by side. Thus when not dealing with profound questions about the nature of reality our forebears would most probably have begun to search for limited kinds of understanding in ways not driven by the desire to achieve anything else at all. When and how this started we are unlikely ever to know. The wonderings of a long-ago Jamie would never be recorded (see p. 44). Yet the tendency to wonder in this way may be as old as humanity.

When we turn to the intellectual transcendent mode the difficulty of finding evidence of absolute origins does not go away. But surely this mode must have come later. Boyer observes that 'primitive numerical verbal expressions invariably refer to specific concrete collections – such as "two fishes" or "two clubs"', which is, of course, characteristic of the construct mode of functioning.[9] Before we can speak confidently of transcendent-mode activity we need some evidence of the systematic manipulation of number – or some other relational system – in the absence of concrete exemplification. We have such evidence in an Egyptian papyrus dating from around 1650 BC and known as the Rhind papyrus, after the collector who brought it to Britain in the nineteenth century, or – more fittingly – as the Ahmes papyrus after the scribe who copied it from a work that he claims was several centuries old.

The papyrus begins with a table expressing the fraction $\frac{2}{n}$ as the sum of unit fractions – that is, fractions with the numerator 1 – for all odd values of n from 5 to 101. Thus $\frac{2}{5}$ is shown to decompose into $\frac{1}{3} + \frac{1}{15}$. And similarly $\frac{2}{201}$ decomposes into $\frac{1}{101} + \frac{1}{202} + \frac{1}{303} + \frac{1}{606}$.

It is not clear why particular decompositions were chosen for tabulation. For example, Boyer asks: Why not decompose $\frac{2}{101}$ into

$\frac{1}{101} + \frac{1}{101}$? He suggests that one of the reasons for decomposing $\frac{2}{n}$ in the ways chosen may have been to finish with fractions smaller than $\frac{1}{n}$. One might wonder if $\frac{2}{n} = \frac{1}{n} + \frac{1}{n}$ was simply too obvious to tabulate. But G. G. Joseph says that, for some unknown reason, it was not permissible in Egyptian computational practice at this period to write $2 \times \frac{1}{2}$ as $\frac{1}{n} + \frac{1}{n}$.

The system was thus odd to our minds, one of its strangest features being the prominence given to $\frac{2}{3}$, which was the exception to the general principle that unit fractions were used exclusively. The standard way of finding one-third of a number was apparently to find two-thirds and then halve the result! Joseph, understandably, calls this perverse. However, he emphasizes, as does Boyer, that whatever the oddities of the system, general rules were known and used. The papyrus gives complicated – to our minds curiously complicated and devious – illustrations of how to apply its procedures to practical problems like dividing seven loaves among ten men. But even if the purposes were practical a level of generality had been reached which justifies us in saying that the intellectual transcendent mode must have been in operation.

Babylonian mathematics at around the same time appears to have been even more advanced – indeed, in many ways highly sophisticated. Most of the texts known to us today are 'Old Babylonian' – that is, they come from somewhere around 1800 to 1600 BC. According to Neugebauer the dating is quite reliable on palaeographic and linguistic grounds.[10] He adds that nothing is known of the historical antecedents of this remarkable period of achievement.

The Babylonian texts are of two main kinds: either they provide mathematical tables or they set mathematical problems. There are, for instance, tables of reciprocals, of squares and square roots, of cubes and cube roots. As for the problem texts, they can be further divided into two classes. One of these states the problem and gives a detailed account of how to solve it, often ending with the words 'such is the procedure'. Problem texts in the other group present a whole series of graded problems, with the solution provided. These were therefore presumably exercises to be worked through by the students. And it is interesting to note that all the problems on a tablet are liable to have the same solution, which tells us once more that what the students had to learn was the method. Some of the expressions in the more difficult problems were horribly complicated. The standard procedures provided step-by-step paths towards simplification.

Babylonian algebra did not use our kind of algebraic notation. Where we would write 'x + y', they might say 'length + width'. At first sight this suggests construct-mode rather than transcendent-mode functioning. However, Neugebauer points out that reality was often flagrantly disregarded. For instance, a problem may be ostensibly about wages to be paid for labour at a given rate per man-day. But it often turns out to be, in fact, a matter of manipulating numbers, as becomes plain when we find that sums and products may sometimes combine the number of men and the number of days! Again, as Neugebauer says: 'It is a lucky accident if the unknown number of workmen, found by solving a quadratic equation, is an integer. Obviously the algebraic relation is the only point of interest, exactly as it is irrelevant for our algebra what the letters may signify.'[11]

Like Neugebauer Boyer also questions the traditionally accepted view that, before the Greek classical period, work with numbers, or mathematics generally, was never a matter of developing theory for its own sake. He believes that sheer intellectual concerns must have been strong in the Old Babylonian period since direct links with practical purpose are often 'far from apparent'. And he too makes the point that words like 'length', 'breadth', 'area' and 'volume' served effectively as 'unknowns' in the modern algebraic sense, since an 'area' might very well be added to a 'volume'. In other words, these terms in the algebraic context had ceased to be used for imaginative constructs based on experiences of the physical world. But they are no doubt traces of earlier construct-mode activity.

Babylonian geometry was relatively weak. But there is enough evidence to establish the existence of a remarkably sophisticated theory of number, together with a high level of competence in algebraic manipulations; and there was a great interest in those sets of numbers which can be used as sides of a right-angled triangle – that is, numbers that satisfy the relation $a^2 + b^2 = c^2$. It was certainly well known that the sum of the squares of the sides of a right-angled triangle equals the square of the hypotenuse. The mathematicians of Old Babylon seem also to have been interested in the ratio of the length of one side to that of the other and to have studied this relationship systematically. Neugebauer thinks that a general formula for the production of 'Pythagorean triples' was probably known to them.

Thus if there is any doubt about Ancient Egypt there seems to be none about Old Babylon: at least 1,500 years before the start of the Christian era the intellectual transcendent mode had definitely arrived upon the human scene.

So we turn finally from the intellectual line of development to the emotional one. The analogy between the two should not be pushed unduly. Nevertheless, it is certainly worth seeing how far it will go without being forced; and so we shall look to see whether evidence can be found not only for *some* emotional development running in parallel to the growth of the intellect but also for a comparable division between 'construct' and 'transcendent'. That is, we shall try to discover whether a value-sensing construct mode and a value-sensing transcendent mode can be distinguished.

The first thing is to be quite clear about how we shall know these modes if we find them, so let me recall their characteristics.

A value-sensing construct mode would be one where the main component of experience was an apprehension of transpersonal importance, powerfully felt, but where the functioning of the mode depended upon the support of the imagination. That is, the imagination would be needed to provide a context within which the mind could operate; and this imagined context would be built up from our ordinary experience of things and events in the world. However, *explanation* of these events would not be the main aim. That is, emotion would have become (relatively) acognitive, as thought can become (relatively) dispassionate – and roughly to the same degree.

Just as thinking in the intellectual modes is by no means entirely unemotional, so an analogue on the emotional side would not be expected to be wholly acognitive. There would certainly be thoughts attendant upon the emotional enterprise. If one responds to a sensed importance, thoughts about the nature of the importance are bound to arise. These would be intrinsic to the undertaking and derivative from it. They would be offshoots of the emotional response. In some circumstances they might be offshoots of an effort to make the appropriate *kind* of emotional response.

As in the intellectual case genuine novelty would be entailed. Specifically it would be true by definition that what is perceived to matter must surpass the personal life. To the extent that it does not, there has been no real movement beyond the line mode. Thus when the main value feeling in religious activity is concern about an afterlife in which there may be reward and punishment for what one does in this life, any attempt to go beyond the line mode has failed. (The counterpart on the intellectual side occurs when line-mode goals intrude so seriously that they distort the processes of thought.) If an activity is to belong in the value-sensing construct mode, then the locus of concern must not lie in the course of one's own life, whether 'here' or 'hereafter'.

Emotion in the line mode stems from personal goals: one's own prosperity and health; what kind of house one has – or car, or jewellery, or clothing; whether one's love is reciprocated; whether one is fairly treated, and so on. Concerns about how one will fare in an afterlife are not radically different, even if we are less naive in providing for these than, say, the Pharaohs.

Having considered, then, the main characteristics of the value-sensing construct mode we can say that the value-sensing transcendent mode will share them, with the crucial exception that the need for a constructed context has gone, so that self-transcending values can now be experienced and responded to without the props provided by the working of the imagination.

These two modes – and especially the transcendent one – are certainly less familiar to us today than their intellectual analogues; and they are less easy to think about and write about. However, once their defining features have been recognized it is not hard to find evidence that they have indeed formed part of the repertoire of at least some minds. It is much harder to say, even approximately, when they first appeared on the human scene. Indeed, the problem is more severe than in the case of the intellect, precisely because emotional response resists systematic recording and codification.

Not surprisingly the construct mode appears to have been the more widely prevalent. The history of religion reveals the frequent appearance of a mode marked by the constructive activity of the imagination. Over and over again we find the divine represented by images drawn from our everyday experience.

Prototypically in Western cultures God has been thought of as an old man with a white beard sitting on a throne in heaven. And heaven has been imagined as a place of light and peace somewhere beyond the sky, while hell and its flames lie far below our feet. Other societies have used similar, though culturally adjusted, images. However, Keith Ward tells us that, in each of the five major religions which he considers, there have been those who have seen the need to leave such concreteness far behind.[12]

Ward's survey of these religions will prove to be of great value for the issues we have to consider. But before we turn to his evidence it may be worthwhile to look briefly at the writings of a renowned – and relatively modern – Christian mystic who has things to say that may help us to appreciate what the value-sensing construct and transcendent modes are like, at least as they functioned for him.[13]

St John of the Cross lived in Spain as a Carmelite friar from 1542 to 1591. His written works were not many, but among them is one called *The Ascent of Mount Carmel*, which John introduces engagingly with the words: 'This treatise explains how to reach divine union quickly.' It then emerges that the proposed way of achieving the aim entails passing through two stages. And these turn out to correspond quite closely to the two value-sensing modes as I have defined them.

There is more to John's method than I shall go on to say; but the part that is relevant here is none the less quite central, and in picking it out I have tried to avoid distortion though I have been unable to avoid omission.

John draws a distinction between the two practices of meditation and contemplation. He returns to this repeatedly, obviously considering it to be essential. Meditation is for 'beginners'. Contemplation is for 'proficients'. Meditation is 'a discursive act built on forms, figures, and images, imagined and fashioned by [the] senses'; and it is 'necessary to beginners that [i.e. in order that] the soul may be enamoured and fed through the senses'. Examples are given of images which may be found helpful: Christ crucified; God seated on a throne with resplendent majesty; a beautiful light, and so on. However, skill in this kind of activity is never to be considered as the goal. It only provides steps to that goal, steps which must be used and left behind like a flight of stairs, or 'we would never reach the level and peaceful room at the top'.

It is clear that meditation in John's sense belongs in the value-sensing construct mode. But in his view this mode, though useful, is strictly limited in its value for us since the divine reality far surpasses anything that we can know through the imagination. So when the beginner has achieved all that discursive meditation can offer John's advice is to try to move on to the much more difficult activity of contemplation. To call this an 'activity', however, may be somewhat misleading. The active part of it consists largely in the rejection of all ordinary experience – sensations, memories, attachments, ideas – with the aim of entering into an essentially passive state of awareness. In this state, John tells us, the soul's work is that of 'attentively loving God and refraining from the desire to see or feel anything'. Likewise, he adds:

If individuals were to desire to consider and understand particular things, however spiritual these things may be, they would hinder the general, limpid and simple light of the spirit. They would be interfering by their

cloudy thoughts. When an obstruction is placed in front of the eyes, one
is impeded from seeing the light and the view before one.

The advice, then, is that, if our imagination is not enough, our
rational powers are certainly not enough either. Indeed, they can
seriously interfere. The recommended way is to empty oneself –
ultimately of everything except loving belief – and then wait for
God to fill the void.

  And how, according to this account, does God fill it? He fills it,
John tells us, with an experience of the divine which is understood
clearly to be ineffable. People who have this experience know
they can never give an adequate account of what they have felt.
John speaks in this context of the 'touches' of God and says:

These touches engender such sweetness and intimate delight in the soul
that one of them would more than compensate for all the trials suffered
in life, even though innumerable ... They are so sensible that they
sometimes cause not only the soul but also the body to tremble. Yet at
other times with a sudden feeling of spiritual delight and refreshment,
and without any trembling, they occur very tranquilly in the spirit.

  John is talking, then, about intense emotional experience. He
speaks also about a kind of knowledge that comes through
contemplation but this is variously described as 'vague', 'dark',
'general' and 'obscure'. It is nothing like the kind of knowing that
comes by way of the intellect. Indeed, as he explains: 'In order to
draw nearer the divine ray the intellect must advance by unknow-
ing rather than by the desire to know, and by blinding itself and
remaining in darkness rather than by opening its eyes.'
  John speaks constantly by paradox, as befits an attempt to
communicate the ineffable. But his account of what happens when
proficiency in contemplation is attained fits closely with the criteria
for the value-sensing transcendent mode. The locus of concern is
certainly not in space-time. Concern centres rather on something
conceived as infinite and eternal – something conceived also as
having supreme value. Accordingly the response evoked is one of
deep emotion. Line-mode goals are specifically excluded from

contribution to this response; indeed, they are seen as interfering with it, as are any of the efforts of human reason.

What we must next ask is whether the kind of experience that John describes is unique or very rare, or whether similar accounts can be found to come from diverse sources in human history. For this purpose, Ward's book is a valuable resource. He surveys five major religions – Judaism, Christianity, Islam, Vedantic Hinduism and Buddhism – looking to see whether, at the core of their orthodoxy, he can find something that they hold in common – something similar in 'the basic structure of faith'. What he claims to find in each of them is, first, a crucially important experience that he calls the 'iconic vision', and, second, a belief that this vision calls forth 'a response of self-transcendence'. The iconic vision is a sense that things point to an ultimate reality beyond themselves – a reality of supreme value that we can recognize, though not always easily.

Ward says that such a sense is best evoked not by science or philosophy but rather by poetry or music; 'it is a major heresy of post-Enlightenment rationalism to try to turn poetry into pseudo-science, to turn the images of religion, whose function is to evoke eternity, into mundane descriptions of improbable facts.'[14] Other writers have argued similarly. Toynbee, for instance, holds that, if we are to talk about mysteries, we must do so indirectly through hints and paradox; and for this we need a means of communication which comes closer to music than to the language of logic.[15]

The method Ward uses for his inquiry is to pick 'impeccably orthodox figures' within his five traditions and look closely at what they said. Ideally one would want to look rather at what they *felt*, but this, of course, is impossible: what we can know is unavoidably mediated by their words. Also they are philosophers, not poets, so the mediation is by way of reflective thought. But Ward's method is probably the best that can be done.

The 'figures' he chooses are Ramanuja and Sankara to represent the Indian Vedantic tradition; Buddhaghosa and Asvaghosa for Buddhism; Maimonides for Judaism; al-Ghazzali for Islam; and Thomas Aquinas for Christianity. In all of them he believes he can find evidence for what he calls the 'dual-aspect doctrine' of the divine. This is the notion that we can know God through manifestations in the ordinary world and yet we can also know that God utterly transcends finite space and time. The iconic vision is 'the discernment of the infinite in and through the finite' – that is, by means of symbols and metaphors. And yet in each tradition there is evidence of an awareness that God is 'truly far beyond our images of him'.

Thus the symbols 'point beyond themselves'. Or, as Ward also puts it, they are a ladder that one must both climb and throw away, 'for the point of the ladder was to bring the mind to a new viewpoint, not to be revered as a sacred idol'.[16] Thus the symbols are there to be used *and transcended*.

These words readily relate to the notion of two modes of apprehending the divine, one dependent on the human imagination and one able to dispense with this support. Also Ward repeatedly makes the point that the iconic vision is not primarily intellectual. At one stage of his argument he does express disagreement with those who think religion consists in 'having certain feelings'. This is because he wants to emphasize that religion is a way of life, aimed at a goal of liberation from the limiting imperfections of human existence – a goal that entails some notion of the nature of the ultimately real so far as this is necessary to explain how these limitations come to be and what can be done about them. On the other hand, he tells us: 'It is most important to grasp the extent to which agnosticism lies at the heart of orthodoxy'; yet there is 'the vivid, direct and intense experience of overwhelming value and reality which comes to some exceptional souls as a vision of God'.[17] This kind of experience is not – could not be – primarily intellectual, or cognitive, certainly not dispassionate. Rather it is experience of 'a reality which is supremely desirable and the source of all other desirable things'. (Ward is speaking at this point of Maimonides, within the Judaic tradition, but as always he is trying to bring out that which is common to the major faiths.) Thus the experience is powerfully emotive; and the doctrines about the nature of reality stand, if Ward is right, in a subsidiary relation to the primal vision and to its significance for life's goals.

They are to be developed 'so far as is necessary', with the acknowledgement always that a full intellectual explication of the divine nature is for ever beyond what our minds can do.

All of this fits well with the idea of two modes of experience that are the emotional analogues of the intellectual construct and the intellectual transcendent modes. It is necessary to add, however, that there is some ambiguity in the way Ward speaks of the dual-aspect doctrine. Sometimes he seems to be talking about two ways in which we can respond to intimations of ultimate reality; and sometimes he seems to be talking about the nature of God. Thus when he discusses the paradox of postulating a God who is both utterly beyond space-time and yet active in the world he says: '. . . one can say that one has two rather different points of view of the same subject.' Yet a little later he speaks of 'two modes in which God exists'.[18] In general, he appears to move between these two positions.

For the present argument the question of whether God exists in two modes – or, indeed, at all – is not one that must be tackled. What matters is whether there is evidence that human beings have experiences of the two distinctive kinds we have been considering. It seems that they do, but not by any means that they all do, especially with regard to the transcendent mode. But then the same is true on the intellectual side. Those competent in logic and mathematics or in any form of highly disembedded reasoning that does not rely heavily on imagery have always been few.

So far we have been concerned to specify the nature of the value-sensing construct and transcendent modes and to see if there are human experiences that fit these categories. Having established that there are and that they are widespread, if relatively uncommon, we must next consider what can be said, with any reasonable confidence, about the measure of their antiquity.

The earliest of the thinkers chosen by Ward to represent the five traditions is the Buddhist Asvaghosa, who lived probably in the second century AD. However, the only complete text of his work 'The Awakening of Faith' that is available to us dates from the sixth century. The other Buddhist, Buddhaghosa, lived in the fifth century, the two Hindu thinkers in the eighth and twelfth centuries, al-Ghazzali in the eleventh, Maimonides in the twelfth, and Aquinas in the thirteenth centuries. But, of course, the origins of the five faiths go back much further; and in the case of Judaism and Hinduism no definitive historical starting-point can be assigned.

Unsurprisingly it seems to be the case that in their earliest forms these great religions were a complex mixture of different components. Consider, for example, the collection of hymns called the Rig Veda, which are the oldest documents in the Hindu tradition. They contain much that belongs in the line mode for they request material prosperity and well-being in a very direct and down-to-earth manner. Also they are concerned with explanations of the cosmos in a way characteristic of the core construct mode. Yet there are glimmerings also of a purer spirituality.

The best impression I have been able to form is that the value-sensing construct and transcendent modes are probably at least as old as their intellectual counterparts, which would place their advent in the middle or earlier half of the second millennium BC. But there is no way to be sure that they were not available, at least to some minds, long before that time. Alexander Marshak, having analysed in great detail the markings found on various ancient bones and pebbles, claims that the marks are neither random nor merely decorative, but that they are notational and were probably used to record phases of the moon. Commenting on the significance of his analyses he argues that, though the Ice Age people cannot be thought of as astronomers or scientists in the modern sense, nevertheless the roots of science lie in their activities.

It must be regarded as possible, though certainly not proven, that all of the modes go back far into prehistory. We have not, and are never likely to have, any way to be sure.

10

The Modes and the Advent of Science

'... though everything comes by antecedents and mediations – and these may always be traced farther and farther back without the mind ever coming to rest – still, we can speak of certain epochs of crucial transition, when the subterranean movements come above ground, and new things are palpably born, and the very face of the earth can be seen to be changing.'

– H. Butterfield

'This new tinge to modern minds is a vehement and passionate interest in the relation of general principles to irreducible and stubborn facts.'

– A. N. Whitehead

We have seen that the construct and transcendent modes do, indeed, have two branches, intellectual and emotional or value-sensing, which can be traced back some way in human history. We cannot be confident about precise beginnings, but for two or three thousand years at least these varieties of human experience have existed. What must now be noted is that, after the beginnings, development does not appear to proceed in a smooth line. A mode once well established in a particular culture may later wane. It will not be lost entirely from the consciousness of certain individuals but it may lose prevalence, it may cease to flourish. Dark Ages may come. And this is as true for the value-sensing modes as for the intellectual ones.

The core varieties of the point, line and construct modes seem to occur universally in some form or other, which is not to say that they develop without powerful social influence. However, once we come to the intellectual and the value-sensing modes, which from now on we shall call the *advanced modes*, then if a mode is not favoured and fostered in a given social group, most young people will not come to it, or at least not to a developed form of it, unaided. Groping towards it will largely fail and be abandoned. This does not mean that human minds are passively formed. Human beings of all ages are active participants in the processes of communication by means of which they grow, so it would be misleading to think of children as being 'moulded' by the adults who rear them (see chapter 6). And yet it is within a society that we become what we are, and whatever we achieve is not achieved alone.

At the same time there are always some individuals who act as influences that promote change. And there are circumstances – physical and economic – that may encourage or even necessitate change, as well as limiting or containing it. Also there are sudden waxings and wanings that can be hard indeed to explain. Tides flow, and we cannot always see the moon.

Recognizing the complexity of such matters, let us return now to a consideration of the relative balance – or imbalance – in the modern Western world between intellectual and emotional development and try to see what has led to the present state of affairs. For this purpose we must go back at least as far as the Middle Ages, before the vast upheaval brought about by the advent in Europe of modern science and technology; and we must ask what then changed. It will be much more a question of 'what changed' than of 'why'. Historical explanation is not the aim. That would be far beyond the scope of this book and of my competence.

Science, as we now know it, is an activity in which three modes combine. They are not fused, they are used in conjunction. Two of these three, the intellectual construct and the intellectual transcendent modes, are able to yield imaginative hypotheses, rigorous reasoning and quantification. But to them must be added, most importantly, that subdivision of the point mode which comes into play when we make precise observations of the kind we call 'objective'. Such observation is a point-mode activity in that the locus of concern is, very strictly, what is happening here and now; but it is a variant in which the close unity of the four original components has been decisively breached. Perceiving and acting are dominant; emotion is, so far as possible, excluded; and even reasoning has, in the process of observation itself, a quite minor part to play.

This at least is the aim. The goal is to protect 'observed fact' from error or distortion of any kind. It has at once to be acknowledged, however, that there are limits to the possibility of arriving at facts, certain and ultimate.

The most obvious source of trouble lies in the imperfection of our senses. On the optician's chart we may read 'Q' as 'O'. Lenses can usually correct this error readily enough, so that we then see the *fact* that the letter is 'Q'. And, in general, instruments can now greatly enhance our senses. What instruments reveal, however, is still not limitless, and it doubtless never can be. Also we may read them wrongly.

Other problems arise that are in some ways more intrinsic to the enterprise, harder to recognize, harder to deal with. A 'fact' is something perceived and consciously noted. For scientific purposes it is something described, recorded. But both perceiving and describing are ineluctably selective. We do not consciously perceive all that is available to our senses; and infinitely many ways of describing what we do perceive are always open to us. Consider a very simple example. If you are shown a jar containing liquid, you may give an account of the height of the column of liquid but neglect to mention the cross-section or many other characteristics of the display. Now for some purposes only the height may matter, for other purposes the neglected aspects may be of crucial importance.

The general point is that one's conception of the question one is trying to answer affects what one looks for and what one reports. It affects what is seen as relevant. Thus theoretical preconceptions and reported observations are by no means independent of one another. Theories – or, indeed, beliefs not conscious enough to be called theories – guide the nature of the observations; and the guiding assumptions are often not recognized as being open to doubt.

It follows that observations are sometimes declared invalid, however carefully they have been made. For example, evidence that homoeopathic remedies are effective in certain illnesses may not be accepted as 'fact' because what the patients have swallowed has been so massively diluted as to contain not a single molecule of the original substance. Now if there exists at the time no theory acceptable to the community of scientists that gives an account of how benefit could possibly be conveyed in the absence of molecules, then the observations will tend not to be given the status of fact at all; this is the case even if they have been made in ways that satisfy all the conditions usually required of evidence. Instead they will be regarded as 'anomalies', not to be taken seriously but rather to be explained away. However, this situation may very rapidly change if someone arrives at a new theory.

It is important to recognize these complexities; but the recognition does not alter the crucial dependence of science on observations made with as little bias as possible and with the aid of the best recording and measuring instruments that we can design and make. We also need our imagination, our rational powers and our mathematics in order to devise hypotheses, plan how to test them and consider what the results imply.[1] All these things must come together in controlled and disciplined ways, as began to happen in dramatic fashion during the sixteenth and seventeenth centuries in Europe.

This coming together was not something wholly new. Its origins can be traced back much further, certainly as far as the Greece of the fifth century BC. The Greeks may themselves have owed much more to earlier cultures than in the past our Euro-centric viewpoint has let us see. However, for much of our modern mentality we do have to acknowledge them as forebears. Among their notable contributions was a conception basic to the whole scientific enterprise: the idea of nature as an enduring, ordered, impersonal system about which it is possible to make inferences. The Greeks also developed deductive reasoning to a very high level, as in their geometry and logic. However, two things hindered them as scientists: they lacked the technology needed for precise measurements; and on the whole they attached relatively little importance to observation. Plato, for instance, held that material things were merely instances of eternal Forms – a doctrine that did not encourage respect for the senses as a means of arriving at universal truths.

The great exception to the undervaluing of observation was Aristotle, who is often regarded as the father of modern science. By his very pre-eminence, however, he became in some ways an obstacle to the progress of science in Europe, as we shall see.

It is very well known that much of the intellectual tradition of Ancient Greece was lost to Europe early in the Christian era and was not recovered until centuries later. The most severe blow came when Islamic invaders overran parts of the Byzantine Empire in the seventh century, breaking the unity of the Mediterranean; but long before that the influx of other invaders had severed the links that came by way of Rome. In spite of this, the West was never wholly cut off from the Greek intellectual tradition. For instance, Pliny's *Natural History* was a vast compilation of Greek learning and observation that served as a textbook throughout the early Middle Ages. And St Augustine was deeply influenced by Platonic views, particularly by the notion of the eternal Forms. There was, indeed, a continuing respect for antiquity that could well be regarded as excessive.

It would be wrong to suppose that the early Middle Ages yielded no original inquiry into natural phenomena, observed in their own right. Bede, in the eighth century, provides a striking exception to the general rule. In 725 he furnished an account of the tides which shows that, in addition to using the standard authorities like Pliny, he used his own observations. His works of reference provided him with the general notion that the tides are caused by the attraction of the moon; but he made his own observations of the effects of the wind and he saw for the first time that, even along the same shore, there are variations that must be tabulated. Crombie says of him:

Though often relying on literary sources when he could have observed with his own eyes – as, for example, in his account of the Roman Wall not ten miles from his monk's cell – Bede never copied without understanding. He tried to reduce all observed occurrences to general laws, and, within the limits of his knowledge, to build up a consistent picture of the universe, tested against the evidence.[2]

A few other examples of what we might admit as science can be found, but they are sparse indeed. What seems to have happened is that the whole of the mental life of Europe had come to be dominated by what Ward calls (though not in this context) 'the iconic vision'. In other words, the culture overwhelmingly favoured and promoted the value-sensing construct mode.[3]

This seems to have been particularly true of the period from the ninth to the twelfth century. During this time the prestige and power of the value-sensing construct mode showed itself in an attitude to nature – and to the understanding of nature – that was deeply inimical to objective observation. The general idea was that to comprehend a natural phenomenon you have to see it as standing for – or, in Ward's terms, 'pointing to' – a reality that lies beyond it. On this view facts are not interesting in their own right, or even for their practical usefulness. They are interesting only to the extent that they can serve as symbols for some religious truth. Crombie gives examples of the way in which this worked:

The moon was the image of the Church reflecting divine light, the wind an image of the spirit, the sapphire bore a resemblance to divine contemplation, and the number eleven, which 'transgressed' ten, representing the commandments, stood itself for sin.[4]

The same kind of habit of mind is exemplified by the bestiaries, in which animals symbolize vices and virtues.

Now the crucial point to grasp – and it is not easy for the twentieth-century mind – is that this was deemed to be as far as one should go in the effort to understand nature. Once symbolic relations like these had been apprehended there were no questions left worth asking. More than this, even: it came to be thought positively improper to press inquiry too far. Nature was conceived as a kind of goddess who wanted to remain veiled – hidden in some measure from our eyes. Curiosity was, then, at risk of becoming impiety and investigation had to be curbed, out of respect. The only safe way was to approach the mystery cautiously and indirectly through allegory and poetry.

Thus there grew up the genre of the 'philosophical myth', often called an 'integument', meaning a kind of covering. The integument was supposed to contain spiritual truth in the form of some secret only hinted at or half-revealed; so that a person 'takes from such a work what is useful according to his own capacity'.[5]

John Scotus Erigena wrote in the ninth century:

Just as poetic art, through untrue stories and allegorical representation, constructs its moral and physical doctrine towards the arousing of the human spirit ... so theology, as if by a certain poetry, shapes Holy Scripture by fictions of the imagination towards the counsel of the soul.[6]

There could be little prospect for the objective point mode, or indeed, for anything approaching modern science, when this was typical of the prevailing spirit of the time.

It was not until the twelfth century that things really began to change. Among the earliest clear signs of this was a revival of the intellectual transcendent mode in the form of mathematics. As Boyer puts it:

At the beginning of the twelfth century no European could expect to be a mathematician or an astronomer, in any real sense, without a good knowledge of Arabic; and Europe, during the earlier part of the twelfth century, could not boast of a mathematician who was not a Moor, a Jew or a Greek. By the end of the century the leading and most original mathematician in the whole world came from Christian Italy.[7]

The critical stimulus for change was the influence of the Arabs as it reached Europe, principally through Sicily and Spain. From around the middle of the eighth century AD various outstanding Arabic centres of learning had been established, the most notable being in Baghdad where Caliph al-Mansur founded the 'House of Wisdom'. This contained a university, an observatory and a large library in which were collected manuscripts from many sources: Babylonian, Indian and Greek.[8]

As the twelfth century progressed more and more of these works were translated first from Arabic into Latin, then also into Spanish or Hebrew. And direct translations from the Greek began to appear. Euclid's *Elements* became available in various versions, as well as the works of Archimedes and Ptolemy and the *Algebra* of al-Khowarizmi (whose name gives us the word 'algorithm'). This must have brought a great new vision of intellectual possibilities to receptive Europeans.

Among such people was an Englishman, Adelard of Bath, who was one of the translators. It is not clear where his knowledge of Arabic was acquired but we know he spent a long time studying in Europe. There exists a fascinating record of a conversation that he had with his more conservative nephew on his return home. The nephew asked: '. . . when you see plants springing up, to what else do you attribute this but to the marvellous effect of the wonderful divine will?' To which kind of argument Adelard replied that, although natural phenomena were the will of the Creator, they were 'not without a natural reason too', and that 'so far as human knowledge has progressed it should be given a hearing. Only when it fails utterly should there be recourse to God.'⁹

This marks a very clear turning-point. The hold of the value-sensing construct mode in European minds would remain strong for centuries more. But the process of challenging its dominance had at last begun.

We have evidence, then, that for many centuries in the history of Europe there was an imbalance in the direction opposite to that which obtains today: the emotional side of development beyond the core modes was cultivated and the intellectual side was relatively neglected.¹⁰ This tells us already that the present state of affairs is not inevitable. But it does not tell us whether one imbalance is just as bad as the other, or whether it would be possible – and, if possible, desirable – to develop a culture where both the intellectual and the emotional would be fostered and valued. For it might be that they are truly inimical to one another. We should also be careful not to conclude that the way in which the value-sensing construct mode manifested itself in the Middle Ages is the only way possible. Within any mode there is scope for growth and change.

The progress towards the establishment of modern science took a long time. It was not until the seventeenth century that this was fully and decisively achieved. However, it is interesting to note that the principles involved in combining the modes were clearly stated by Robert Grosseteste, who was born around 1170. Grosseteste was a distinguished scholar who became Chancellor of Oxford University – perhaps its first Chancellor – and later Bishop of Lincoln. His exposition was surprisingly sophisticated and thorough, stressing the importance of observation and experiment, of imaginative leaps, of classification and of the abstraction of general principles. He also recognized that the certainty attainable in mathematics is not attainable in science, where one can never know all the possibilities.

Grosseteste was very influential in his day, and subsequently. Yet it is one thing to enunciate principles and another to apply them. The full effect of the application of Grosseteste's principles was slow to come.

A cultural legacy from a more developed to a less developed society gives rise to a major problem: the recipient is, by definition, not well placed to judge the worth of the bequest in a discriminating way. If everything glitters, the heir to the treasure may not be able to tell the gold-dust from the yellow sand. Neither the Greek nor the Arab tradition as inherited by medieval Europe turned out to be pure gold. But it took some time for the distinctions to be drawn.

One of the main difficulties lay in recognizing those respects in which Aristotle was misleading. He had opened up the road to science, and yet he blocked it at the same time. For one thing, he was more given to classification than to measurement – less of a mathematician than, say, Plato or Pythagoras who, as Whitehead points out, stand in some ways nearer to modern science.[11] Aristotelian Logic became very influential in the later Middle

Ages, but it was not the kind of transcendent-mode activity that science most needed at that time. Also some of Aristotle's specific ideas about the physical world were distinctly unhelpful. These came to have a very strong hold and it was not easy to get them out of the way.

Aristotle's theory of motion is the prime example. It lay at the core of his philosophy of science and of his cosmology, which systems, though by no means unchallenged, dominated European thought from the thirteenth to the seventeenth century. For Aristotle 'motion' meant any kind of change. Change of position – or local motion – was one kind among others. The central idea was that all things are moved by something, whether by the fulfilment of some intrinsic principle natural to them, as in the case of living things, or by some external force. Also movement was held to cease as soon as the activity of the 'mover' ceased. The universe was conceived as a number of celestial spheres in which there was a prime mover, or original source of motion; and to each sphere was assigned some kind of 'soul'.

Butterfield points out that the postulation of unseen 'Intelligences' whose function is to impart movement is positively invited by Aristotle's starting-point: the assumption that what principally needs to be explained is movement rather than the absence of movement or the state of rest. Such a theory, by leaving 'the door half-way open for spirits', as Butterfield puts it, sat comfortably with the habitual modes of the medieval mind. But it was fundamentally unsatisfactory. There were many easily observable phenomena that it quite failed to explain.[12]

In such a case more observation, however important, is not enough. What is needed is a new conceptual framework. And the provision of this is the work of the imagination. As Butterfield says: 'We do not in real life have perfectly spherical balls moving on perfectly smooth horizontal planes – the trick lay in the fact that it occurred to Galileo to imagine these.'[13] Galileo's achievement is made all the more striking by the fact that it came at a time when lubrication was much poorer than now and when well-machined surfaces of the kind we have today were unknown.

His new conception led directly to the idea that movement is something which, instead of having to be perpetually caused to continue, will continue – once started – until it is caused to stop. This was a breakthrough which did much to prepare the way for Newton; and it is a fine example of the intellectual construct mode at work.

The whole story illustrates the important point that it greatly matters which images you use in the constructive activity. Aristotle, when thinking of motion, was using the intellectual construct mode too. But he seems to have had in mind the image of a horse drawing a cart, which was of limited value. Progress in understanding often depends on the development of a new and more appropriate kind of imaginative construction to replace the old. When the old has become habitual, this is one of the hardest things for the mind to do.

Galileo's work seems recognizable to the modern mind as modern science. The troubles that he had when the Holy Office objected to some of his theories are well known; and these two facts may lead us to think that by the beginning of the seventeenth century religion and science were already clearly distinguished and set apart one from the other. We may also be tempted to conclude that the line between science and magic had by this time been firmly drawn. After all, Newton was born in 1642, the year of Galileo's death, and the Royal Society was founded in 1660. However, we would be wrong to suppose that science was generally thought of then just as we think of it now.

Part of the Arabic inheritance that began to reach Europe in the twelfth century was an intense interest in magic, alchemy and astrology. This entailed an attitude to nature that stood in sharp contrast to the earlier European respect for the right of the goddess to remain veiled. A different aim appeared: that of removing the veil. The new guiding question was: how can we find out those secrets that will give us power?

This question is familiar enough to us now. But the common belief then was that the secrets were to be revealed by what we would regard as magical means; or at any rate no sharp distinction was generally recognized between rational inquiry and the occult.

For instance, al-Kindi, a ninth century Arab scholar who made a considerable contribution to the study of optics and perspective, also wrote a work called *On Stellar Rays, or The Theory of the Magic Art*, in which he claimed that the stars, and also the human mind by means of magic utterances, could exert certain influences that were ultimately attributable to 'celestial harmony'.[14]

Ideas of this kind have proved to be very enduring. Certainly they were prevalent throughout the later Middle Ages in some of the most scholarly minds of the times. But it is particularly interesting to discover the strength of the magical tradition among such people as late as the seventeenth century. Frances Yates, in her book *The Rosicrucian Enlightenment*, tells a fascinating detective story about her inquiries into the nature and the influence of the Rosicrucian movement.[15]

Her evidence is far too detailed to summarize here, but the outline of the story can be briefly told. In the Palatinate on the Rhine in the second decade of the seventeenth century there were published two documents known as the *Rosicrucian Manifestos*. These declared the existence of a secret Brotherhood, explicitly Christian and devoted to the furtherance of human welfare. This Order was said to have been started by one Christian Rosencreutz, who was born in 1378 and lived for 106 years. Brother R. C., as he was called, had travelled widely in the East and had there learned the 'Magia and Cabala'. He had used these to enhance his Christian faith, and had also studied mathematics and the making of instruments. He had then begun to recruit and train followers. These now were said to have a centre called the 'House of the Holy Spirit', where they met annually. The rest of the time they travelled around in various countries, incognito or 'invisibly', healing the sick but accepting no payment. The time had now come for the Rosicrucians to initiate, through the power of their enlightenment, the general reform of human society. The manifestos proclaimed, as Yates puts it:

... the discovery of a new, or rather new-old, philosophy, primarily alchemical and related to medicine and healing, but also concerned with number and geometry and with the production of mechanical marvels. It represents, not only an advancement of learning, but above all an illumination of a religious and spiritual nature.[16]

Accordingly, the Rosicrucian alchemy had nothing to do with making gold but rather with the riches of the spirit. Thus the first manifesto declared that Brother R. C. 'doth not rejoice that he can make gold but is glad that he seeth the Heavens open, and the angels of God ascending and descending, and his name written in the Book of Life'.

Yates argues that, most probably, the Rosicrucian Order never actually existed. She believes the aims of the manifestos were religious and political, being associated with evangelical Protestantism and with attempts to bring about the fall of the House of Habsburg. What is significant for our present purposes, however, is, first, the intimate linking of religious fervour with mathematics, mechanics and magic; and, second, the influence of the movement on what followed.

The attempt to bring down the House of Habsburg failed when Frederick, the Elector Palatine, was defeated by the Catholic armies at the Battle of the White Mountain in 1620 and the horrors of the Thirty Years War began. There followed a systematic attempt by the victors, organized probably by Jesuits, to discredit the Rosicrucian movement. And one of the main methods was to accuse the Order and its supporters of being evil magicians deriving power not from angels, as they claimed, but from the devil.

The fact that the manifestos spoke of the Brothers as travelling around in such a way as not to be easily recognized made the accusation all the more alarming. After all, they might be anywhere. They might even be truly invisible!

Yates reports a claim by Gabriel Naudé that in Paris in 1623 there appeared placards announcing: 'We, being deputies of the principal College of the Brothers of the Rose Cross, are making a visible and invisible stay in this city through the Grace of the Most High . . .'[17]

Also she tells how an anonymous work in the same year reported on the placards under the title: 'Horrible Pacts Made between the Devil and the Pretended Invisible Ones'.

Details were given of a meeting of the Order at which a prince of the infernal cohorts had appeared. Before him the Rosicrucians were said to have sworn to abjure Christianity and to have been promised all manner of magic powers, including gifts that would cause them to be admired by the learned and to seem wiser than the prophets.

The period, we must remember, was one of fearful witch-hunts and this was scary stuff. Before then the original Rosicrucian manifestos had apparently been making a great impression in France. Naudé is quoted by Yates as saying that, after the 'novelties' which had surprised the previous generation, such as the discovery of new worlds, and many inventions like the cannon, compasses and clocks, a new age of enlightenment was generally believed to be at hand.

Naudé was careful not to seem to approve of the Rosicrucians, but later he wrote another book in which he distinguished divine magic, religious magic, natural magic (that is, natural science) and evil magic or witchcraft.[18] In this work he suggests one reason why great men may be wrongly accused of witchcraft, namely that they tend to study mathematics and that the marvels produced by mathematics and mechanics seem like magic to the ignorant. In this context he refers to John Dee, the great mathematician and magus of Elizabethan England who, in the opinion of Yates, was a main source of inspiration for the entire Rosicrucian movement.

This explicit reference to a tendency to associate mathematics and mechanics with witchcraft is of great interest. What it means is that men of learning were in real danger; and it must have become crucial for them to dissociate themselves publicly from magic in every way possible.

Even Descartes had to do this. At the time of the famous dreams which so influenced his intellectual development, Descartes was living near the Danube; and it seems he had heard about the Rosicrucians with their promises of a new dawn of wisdom and learning and had been sufficiently intrigued to try to get in touch with them, although, like everyone else, he failed to find them.

However, the upshot was that when he returned to Paris in 1623, at the height of the rumours of Rosicrucian witchcraft, he was suspected of being one of them. And, according to his biographer, Baillet, Descartes's habit of living in solitary retreat even encouraged the notion that he might have become one of the Invisibles. Apparently he managed to dispel this suspicion by the simple device of appearing again in society.[19]

I have to say that I do not find this wholly convincing, since it seems too easy a way of dispelling such a rumour. After all an invisible magician can presumably make himself visible when he chooses! However, the story tells us a good deal about the atmosphere of the times.

Yates believes that, in England, Francis Bacon may also have found himself in a threatened position. She argues that to propagate the advancement of learning during the reign of James I was hazardous. Queen Elizabeth had supported John Dee and his researches. James, on the contrary, had a deep fear of magic and witchcraft. Yates even suggests that Bacon may have played down the role of mathematics in science because he was afraid of the consequences of being associated with magic. It is certainly possible that he was a secret admirer of the Rosicrucians, for after his death in 1626 there was published a work of his entitled *The New Atlantis*, in which he sets out his idea of the perfect society. Yates shows that the inhabitants of this Utopia strikingly resemble the Brothers of the Rose Cross as described in the original manifestos.

The danger of being accused of witchcraft had not entirely vanished by the time the Royal Society came into being. Yates traces many links between the founders and the Rosicrucian movement, or at least its characteristic aspirations. And she says: 'The rule that religious matters were not to be discussed at the meetings, only scientific problems, must have seemed a wise precaution . . .'[20]

Comenius, in congratulating the members of the new Society, called them 'illuminati', a term that had been regularly used of the Rosicrucian Brethren. At the same time he expressed unease about the new and more restrictive policy of confining efforts to the natural sciences for themselves alone instead of making them part

of a movement for religious and spiritual regeneration. He warned
that the work might prove to be 'a Babylon turned upside down,
building not towards heaven, but towards earth'.[21]

If it is true, as I have suggested, that in Europe from the ninth to
the twelfth century the value-sensing construct mode enjoyed
considerable prestige, what are we now to say of prevalent modes
in the centuries that followed, up to the seventeenth century?

We have seen that there was a persistent desire during this time
to discover the secrets of nature and so gain power. But this was
associated with equally persistent avowals of deep religious reverence. No doubt the avowals were not always sincere. No doubt
also many of them were. The idea became widespread that human
beings stand in a special place between the purely material world
and the spiritual world of the angels and of God, a place which gives
them both privilege and duty. As Crombie says, the effect was to
lay emphasis on the sacramental aspect of scientific activities. And
he quotes the thirteenth-century writer, Vincent of Beauvais: 'I am
moved with spiritual sweetness towards the Creator and Ruler of
this world, because I follow Him with greater veneration and reverence, when I behold the magnitude and beauty and permanence of
His creation.'[22] Crombie remarks that this might equally well
have been said by many another scientific writer of the time.

At the other end of the period there is the evidence already
mentioned about Bacon and the founders of the Royal Society. It
appears that Newton, too, was deeply religious and, like so many
of his contemporaries, profoundly interested in alchemy. Indeed,
we now know that alchemical studies occupied him for at least
thirty years, a far longer time than he spent on anything else. He
saw his researches – all of them – as a way of trying to explain the
workings of God in the created universe.[23] The crucial contrast
between this view and the beliefs of the early Middle Ages was
that God was now supposed to want us to try to understand.

What, then, do we conclude as to the modes that were
dominant? Are we to say that the intellectual and the emotional
modes were distinguished from one another and coexisting

harmoniously? Or do we rather have to recognize a version of the core construct mode in which explanation is gradually becoming more sophisticated and powerful?

The latter seems to be the more reasonable account, at least from the thirteenth to the sixteenth century. We find throughout this time little tendency to draw boundaries between the work of understanding nature and the religious response to it.

All attempts to characterize the *Zeitgeist*, especially over such long periods, are open to challenge by counter-example. For instance, there is Montaigne. He urges us to try to find out as much as we can about the laws of nature through our own experience, and there is not much trace of religious emotion in the way he writes about this. He is speaking, of course, mainly about knowing one's own nature, not about studying the physical world.

It remains true, I think, in spite of some exceptions that the general tendency during this period was not to separate attempts to understand nature from religious reverence for its creator. Mystical response was not seen as being in sharp contrast with the use of evidence or the efforts of reason. Copernicus, for instance, 'rises to lyricism and almost to worship' when he writes about the central position of the sun.[24]

But in the seventeenth century there did come something new. As Paolo Rossi says, in spite of all the links with the past there was a genuine discontinuity, a radical change.[25] To recognize what Bacon, Copernicus, Descartes and Newton derived from their forebears is one thing. To recognize in their work the signs of the novelty that was dawning is another. And what was dawning – though it could not be achieved all at once – was precisely the setting apart of the intellectual modes from all others and the practice of conjoining them with one another and with the objective point mode in an astonishingly powerful way. To quote Rossi:

. . . in the case of the history of science – at least from the age of Galileo and quite apart from what was happening in the world of magic – it is justifiable to speak of theories that are more or less rigorous, have greater or less explanatory and/or predictive power, and are verifiable to a greater or lesser degree.[26]

Rossi cites no fewer than fourteen circumstances that contributed to the scientific revolution, among them such well-known facts as the development of more precise measuring instruments and the discovery of new parts of the earth, with the resultant disturbance of old conceptions. But perhaps most important was the gradual realization – and it had to be gradual – that the way to a better understanding of the natural world is not by the sudden revealing of a secret mystery but by the slow accumulation of verifiable knowledge. This realization leads directly to the publica- tion of findings, to the establishment of open critical debate and to the notion of the community of scientists that we entertain today.

This still leaves us with a crucially important question that is usually neglected: what happened to the other advanced modes as the new scientific alliance of modes became firmly established and increasingly strong?

First, however, this chapter needs a brief postscript. It is essential to recognize in discussing science as it has developed since the seventeenth century, how far what actually goes on often falls short of the idealized version. It is undoubtedly true, for instance, that scientific work is often 'messy'; that the instrumentation and the kinds of experiment done depend on local conditions; and that excitement in a lab is often caused not by some revela- tion of truth but by the glimpse of an opportunity to get a research grant or beat one's competitors. That is merely to say, as I have said before, that the intellectual modes are hard to achieve and harder to sustain and that the line and core construct modes are always apt to reassert themselves.

A valuable account of how science really functions in practice much of the time is given by Barry Barnes.[27] He is particularly interesting when he writes about the importance of 'Who says so?' (the question of the credibility of observers and the standing of individuals within the scientific community); and also of 'What counts?' (what is to be taken seriously as new evidence that needs explaining).

Barnes's general argument is balanced and reasonable. Science, he says, is theoretical through and through and it is provisional. Scientific knowledge is what we are 'content to use for the time being as the basis of our understanding of nature'. This does not mean, however, that we should cease to trust and use it; on the contrary it is 'just the knowledge we have found most trustworthy in use'. But at the same time we should not suppose that it is *final* truth, immune to change.[28]

The Advanced Modes after the Enlightenment

'May God us keep
From single vision and Newton's sleep.'
– William Blake

One might have supposed that the independent development of the intellectual modes from the seventeenth century onwards would have entailed a straightforward splitting of the core construct mode of such a kind as to favour a parallel development on the value-sensing side. This did not happen. On reflection it is not surprising.

Religion had permeated very nearly the whole cultural life of Europe for many centuries. Almost every intellectual argument had been related to religious dogma, if not explicitly then at least by means of a powerful background of doctrinal assumptions. Also the Church had been able to impose its views about the nature of the physical universe by censoring theories which it found displeasing. It is very well known that Galileo was tried and condemned in 1633 by a tribunal of the Holy Office, ostensibly for advocating the idea that the earth revolves around the sun. Pietro Redondi argues that the real reason had nothing to do with heliocentrism but lay rather in the much more profound fear that Galileo's revival of atomistic theories about the nature of matter threatened the central doctrine of the Eucharist: the transubstantiation of the bread and the wine.[1] Whatever the reason, the point is that the Church was powerful enough to force Galileo to recant, and to imprison him. And yet he was a man of immense prestige and influence, much respected and admired.

It is less well known that in 1633, hearing of Galileo's fate,

Descartes abandoned plans to publish a book called *The World* which expressed belief in heliocentrism and in an infinite universe. Descartes had no inclination towards martyrdom. Also he was convinced of the existence of God for he did not see how we could have got the idea of an infinite, perfect being except from that being. But in spite of his caution and his genuine belief he did not permanently escape the charge of being irreligious. In 1643 he was summoned to appear before the magistrates at Utrecht and was saved only by the support of the French ambassador and of the prince of Orange. Later his works were proscribed for a while in various universities.

We have seen, too, how great were the risks throughout Europe at this time for any person charged with being in league with the devil. And we have seen that intellectual activity, particularly mathematics, might be enough to cause one to be suspected. Thus, in the relations between religion and science, religion had the upper hand. Giving this up was not going to be easy.

From the mid-seventeenth century onwards science could not be stopped. Indeed, its power was bound to grow at an increasing rate. Humanity had at last set out on a new path to knowledge and the exploration thus made possible was deeply satisfying. Also it soon became evident that science was immensely useful in practical ways.

The realization that momentous change had come, and would surely increase, might well have been restricted for some time to those who were themselves contributing actively to the scientific discoveries. That this did not happen was due to a new phenomenon: the appearance of writers who popularized science and who used results to construct a new world-view. The centre for this kind of activity was France, partly because of the immense influence there of the philosophy of Descartes.

Descartes developed a 'method of doubt' which he described as follows: 'I thought . . . that I ought to reject as absolutely false all opinions in regard to which I could suppose the least ground for doubt, in order to ascertain whether after that there remained aught in my belief that was wholly indubitable.'[2] The idea was to find a sure foundation, and then to build upon it a system of certain knowledge.

For Descartes himself, the existence of God was part of this system. He drew a sharp distinction between mind – or soul – and matter. So far as matter was concerned – and matter included the human body – he believed that we should apply our reason to the full in the search for mechanical explanations; and he saw no conflict between this and the attribution of the origin of all things to God. But those who followed found it easy to adopt his ideas on method and human reason while abandoning his theology.

Fontenelle, one of the first great popularizers of science, spoke of Descartes as one 'to whom might legitimately be accorded the glory of having established a new art of reasoning'.[3] Fontenelle was secretary to the Académie des Sciences for more than forty years at the beginning of the eighteenth century and thus in a good position to know what was going on. He had an easy, readable, witty style. And the rise of a prosperous middle class gave him readers who were free of the old traditions of scholarship and ready for visions of the future. It was not hard to persuade the intellectuals of the new bourgeoisie that methodical doubt and questioning should be preferred to authority as sources of belief. So society became progressively, and in the end very thoroughly, secularized. Nor was this trend by any means confined to France. Much the same kind of thing was happening in England and in the Netherlands; and the tendency soon spread throughout Northern Europe.[4]

Faced with all this, what were the Churches to do? One course of action would have been to redefine their role in such a way as to recognize where they could not compete with science – and where science could not compete with them. This would have amounted to a shift of emphasis towards the value-sensing modes. But it was not the way that was chosen – or even, I think, considered – at least in the Protestant churches, where the response to the Age of Reason can best be described as one of emulation. Religion became more, not less, intellectual. Emotion was played down, contempt for what was called 'enthusiasm' in religion was widespread, and there developed what Butterfield calls 'Protestantism married to the rationalizing movement' – a cold and rather bleak affair.

Thus ascendancy passed more and more completely to the intellectual modes. There were, of course, individual exceptions, as always. For instance, William Law argued passionately against reason and for direct spiritual experience as the only true way to know the love of God within us. He writes to one correspondent:

There is hardly anything more hurtful to true spirituality (the life of God in the soul) than a talkative, inquisitive, active, busy, reasoning spirit, that is always at work with its own ideas, and never so content as when talking, hearing, or writing upon points, distinctions, and definitions of religious doctrines . . .[5]

It is probably true to say that in the eighteenth and nineteenth centuries there developed a tendency for spiritual feeling to express itself by means of music and literature rather than through traditional kinds of piety. In music the outstanding example is Beethoven. Most people who care for his later music recognize in it an intensity of self-transcending emotion that has rarely, if ever, been expressed with such power.

Among poets one of the most mystical is Wordsworth, for whom the great stimulus to spiritual feeling lay in the natural world, and especially in sky, water, rocks and hills. In *The Prelude* he speaks of a childhood filled with wonder and joy that came from his experiences of these aspects of nature – experiences that entailed a strong sense of affinity and communion.[6] There are so many illustrative quotations that it is hard to choose, but consider, for instance, the following:

> . . . for I would walk alone,
> Under the quiet stars, and at that time
> Have felt whate'er there is of power in sound
> To breathe an elevated mood, by form
> Or image unprofaned; and I would stand,
> If the night blackened with a coming storm,
> Beneath some rock, listening to notes that are
> The ghostly language of the ancient earth,

Or make their dim abode in distant winds.
Thence did I drink the visionary power;
And deem not profitless those fleeting moods
Of shadowy exultation: not for this,
That they are kindred to our purer mind
And intellectual life; but that the soul,
Remembering how she felt, but what she felt
Remembering not, retains an obscure sense
Of possible sublimity . . .

Notice that Wordsworth feels it necessary here to justify the
'shadowy exultation' as 'not profitless' – and this because the sense
of 'possible sublimity' was to be valued in itself, rather than for
any reasons having to do with the intellect. The fact that he
thought this worth saying tells us clearly how dominant the
intellect had become.

There can be no doubt that Wordsworth looked on the sublime
feelings as being religious in nature. For instance he speaks of a
'holy calm' that came upon him and of

. . . the spirit of religious love
In which I walked with Nature.

He is giving us in this poem a passionate account of his own
experiences of the value-sensing modes – especially the transcendent mode – and a plea that their importance should be defended
against the encroachments of the intellect. In a famous passage he
speaks of science

Not as our glory and our absolute boast,
But as a succedaneum, and a prop
To our infirmity.

And he calls it

. . . that false secondary power
By which we multiply distinctions, then
Deem that our puny boundaries are things
That we perceive, and not that we have made.

Wordsworth's writings were influential; but they did not redress the balance. By the nineteenth century science was overwhelmingly successful. It worked. And increasing numbers of people came to accept that its success implied the truth of the materialist notions on which its theories were founded. Thus even religious people were for the most part persuaded that inanimate nature was just a huge machine, having been set in motion long ago by God and then left to run according to divine decree. However, so long as this image of the universe could be deemed inappropriate to human beings, the Churches' situation was not entirely desperate. It was still possible to maintain that, God having created us in his own image, we occupy a uniquely important place in the scheme of things.

This last refuge of the believer, as it must have seemed, was decisively attacked when Darwin published *The Origin of Species*, and then *The Descent of Man* and *The Expression of the Emotions in Man and Animals*. Darwin's readers were presented with a mass of evidence for the claim that we evolved from lower creatures by a combination of chance fluctuation and automatic processes whereby some individuals were eliminated and some survived. Any trace of a notion that we were here to serve a divine purpose was absent.

It is no wonder, then, that the old desire to ban the unacceptable should have flared up once more, to the point where the teaching of evolutionary theory was forbidden by law in some parts of the United States where fundamentalist religion was strong. When I taught in a college in Tennessee in the early 1960s this was, of course, no longer so; and yet I was warned that, as far as possible, I would be wise to avoid mention in my lectures of our kinship with monkeys.

It is interesting to note the effect on Darwin himself of the development of his theory. The following quotation is from his autobiography:

In my Journal I wrote that whilst standing in the midst of the grandeur of a Brazilian forest '. . . it is not possible to give an adequate idea of the higher feelings of wonder, admiration and devotion which fill and elevate the mind.' I well remember my conviction that there is more in man than the mere breath of his body.[7]

Then he adds, sadly: 'But now the grandest scenes would not cause any such convictions and feelings to rise in my mind. It may truly be said that I am like a man who has become colour-blind . . .'

Disbelief in Christianity crept over him, he tells us, 'at a very slow rate but was at last complete'. He still did not find it easy to accept that 'this immense and wonderful universe [could] be the result of blind chance or necessity'; and he acknowledged in himself tendencies to theism. But in the end he decided that the mystery of the beginning of all things is beyond us and that he must remain an agnostic.

Along with the loss of religious belief there developed in Darwin what he himself calls a 'curious and lamentable loss of the higher aesthetic tastes'. His mind became 'a kind of machine for grinding general laws out of large collections of facts', so that he ceased to enjoy poetry, paintings or music. This, he tells us, is 'a loss of happiness, and may possibly be injurious to the intellect, and more probably to the moral character, by enfeebling the emotional part of our nature'.

Thus he was well aware of a growing imbalance in himself like that which was affecting the whole culture. The intellectual modes were in the ascendant, the advanced value-sensing ones were on the wane. Darwin contributed to this – and suffered from it at the same time.

He was, of course, by no means the first evolutionist. Indeed, most of his ideas can be found in earlier writings.[8] But he brought them together and supported them by unusually detailed observations. Also the world was finally ready for them. The theories completed the expansion of mechanistic ideas that had been going on steadily. They seemed like the ultimate in enlightenment, the final escape from ancient superstition. We were machines like the rest of nature. We had better face the truth and delude ourselves no longer.

Most theories which achieve great fame are much simplified in the process of being disseminated, and Darwin's ideas about evolution are no exception. It is important to recognize that he entered many caveats about the inapplicability of the processes of natural selection to the development of human societies. For example:

The bravest men, who were always willing to come to the front in war, and who freely risked their lives for others, would on an average perish in larger numbers than other men. Therefore it seems hardly probable, that the number of men gifted with such virtues . . . could be increased through natural selection, that is, by the survival of the fittest . . .[9]

Or again:

We civilized . . . men do our utmost to check the process of elimination: we build asylums . . . we institute poor laws . . . [and we could not do otherwise] without deterioration in the noblest part of our nature.

Or again later:

We must remember that progress is no invariable rule. It is very difficult to say why one civilized nation rises, becomes more powerful, and spreads more widely, than another; or why the same nation progresses more quickly at one time than at another.

The truth is that people often take from a theory what suits them: what excites them, or takes their fancy, or fits well with their own ideas. Thus behaviourist psychologists seized on the idea of chance variations, some of which succeeded while some 'failed'. In behaviourist theories the chance variations were provided by 'responses' that were randomly emitted until some were strengthened by positive reward and so were learned, while others, in the absence of such reward, were extinguished and died away.

The parallel with Darwinian natural selection is obvious; but the fascinating thing to observe now is the enthusiasm with which this doctrine was proclaimed. The behaviourists themselves were filled with passionate purpose. They wanted to be 'scientists', for the name carried great prestige. This prestige had been won by the physical sciences, which seemed to derive their strength, even the very possibility of their success, from mechanistic assumptions about what they studied – that is, about 'matter' as then conceived. Accordingly the new scientific psychology would best prosper, it appeared, in the same way. There was little acknowledgement that what mattered was scientific method; and there was still less recognition that the commitment to particular assumptions as to the nature of the object of study was itself profoundly unscientific and could only encourage the intrusion of distorting emotion.

Behaviourism was never by any means the whole of psychology, but it was very powerful and very arrogant in the first half of the twentieth century. Yet in 1905 Einstein had begun to publish his work on the theory of relativity, and the old deterministic physics based on our ordinary conceptions of absolute space and time was beginning to be replaced by something astonishingly different and new.

Then, following upon the work of Max Planck from which emerged quantum mechanics, earlier conceptions of matter itself – those lying at the very basis of the 'materialistic' theories of nature – were shown to be mistaken. Matter was found not to consist of small solid things that moved around like billiard balls but rather to be basically insubstantial. Its fundamental particles turned out not to be 'particles' in an ordinary sense at all but curious 'things' that had also the properties of waves.[10]

It was discovered that, as energy was given off within an atom, the sub-atomic particles lost that energy discontinuously in packets or quanta of specifiable size. In other words, they moved by 'jumps' between separate pathways; and while they were 'jumping' they could not be considered as continuing to exist. They vanished and reappeared again in what Whitehead calls 'discontinuous realizations', which makes it doubtful whether we may legitimately speak of 'them' at all, in any ordinary linguistic sense. It is, to quote Whitehead again, 'as though an automobile moving at an average rate of thirty miles an hour along a road did not traverse the road continuously, but appeared successively at the successive milestones, remaining for two minutes at each milestone'.[11] Can we then reasonably ask where 'it' was in the meantime – or does this question cease to make any sense? If it makes no sense, the whole basis for traditional materialism has gone.

Add to this now a further problem and the extent of the collapse of earlier notions becomes plain. It has been realized that, at a sub-atomic level, the very act of observation dramatically changes that which is being observed. This is not just because of practical defects in our present measurement techniques. It appears to be a fundamental limitation imposed by the very nature of sub-atomic happenings. Davies puts it thus:

The indeterminacy of the microworld is not just a consequence of our ignorance . . . but is absolute. We are not merely presented with a choice of alternatives, such as the heads/tails unpredictability of daily life, but a genuine hybrid of the two. Until we make a definite observation of the world it is meaningless to ascribe to it a definite reality (or even various alternatives), for it is a superposition of different worlds.[12]

The notion that the act of observation is what causes this superposition of possibilities to collapse into reality seems to be required for the interpretation of certain highly paradoxical and perplexing experimental findings. The implication is that the observer has a quite fundamental role to play; and this has been recognized by some of the most distinguished physicists of the twentieth century. For example, Eugene Wigner tells us that 'it was not possible to formulate the laws of quantum mechanics in a fully consistent way without reference to the consciousness'.[13] So we have the strange state of affairs that, while prominent schools of psychology were trying to push consciousness out of their theorizing, physicists were finding they had to bring it in!

Recent developments in physics have many bizarre consequences that defeat the efforts of the human imagination. We find ourselves in a strange world where 'intuition deserts us, and seemingly absurd or miraculous events can occur'.[14] In other words, the intellectual construct mode reaches its limits and the extension of knowledge depends ever more heavily upon mathematical reasoning.

This helps to explain why the new physics is much harder to

popularize than the old.[15] Most people cannot do without the help of the construct mode – not, at least, given our present ways of educating them. Also the old upbeat optimism of the eighteenth and nineteenth centuries was attractive in ways that cannot be matched by the new uncertainties. We can scarcely now suppose that in a short time all will become clear. No doubt the new puzzles are compellingly attractive to those equipped to consider them. But this attraction is not one that can be widely felt. Thus superseded materialism persists in many minds, for want of any sense of the freshly revealed mysteries.

But what is made of the mysteries by those who do become aware? Does there arise in such minds a renewal of religion? Certainly this can happen. For instance, the extraordinary precision that seems to have characterized the first seconds of the Big Bang is taken by some to show that the universe is planned – and executed – with awe-inspiring power and skill. But conclusive proof of the existence of a Planner remains as elusive as ever.

On the other hand, one thing is clear: the activities of science can no longer reasonably be taken as tending to the conclusion that the universe is a mechanism which we can expect soon to understand completely. That idea is dead.

Thus while science has gone from strength to strength, fulfilling its promises of power, it has also grown humbler. And the way is now open for a general recognition that the value-sensing modes need not compete with the intellectual modes but can properly function in their own way.

As the twentieth century opened, William James delivered the Gifford Lectures in the University of Edinburgh. His title was *The Varieties of Religious Experience*; and few people have ever shown a clearer understanding than he did of the nature of spirituality, and of what it is to have value feelings that transcend the personal life.

His conception of science belonged naturally to the century that had ended:

Our solar system, with its harmonies, is seen now [by science] as but one passing case of a certain sort of moving equilibrium in the heavens, realized by a local accident in an appalling wilderness of worlds where no life can exist . . . The Darwinian notion of chance production, and subsequent destruction, speedy or deferred, applies to the largest as well as to the smallest facts. It is impossible, in the present temper of the scientific imagination, to find in the driftings of cosmic atoms . . . anything but a kind of aimless weather, doing and undoing, achieving no proper history, and leaving no result.[16]

But James was too great a psychologist and too independent a thinker to be bemused by such notions, even in the days before Einstein and Max Planck. 'Humbug is humbug,' he tells us firmly, 'even though it bear the scientific name, and the total expression of human experience, as I view it objectively, invincibly urges me beyond the narrow "scientific" bounds.' James recognized that science and religion 'are both of them genuine keys for unlocking the world's treasure-house'; that 'neither is exhaustive or exclusive of the other's simultaneous use'; and that religion at its higher flights is 'infinitely passionate'. As to the justification of religious experience, he tells us that some people '*know*; for they have actually *felt* the higher powers'.

I am reminded of Charles Darwin's story of an old lady who said to Darwin's father: 'Doctor, I know that sugar is sweet in my mouth, and I know that my Redeemer liveth.' Darwin remarks that the argument is unanswerable. But it is scarcely an argument at all. Rather it is an account of an experience.

I once asked Winifred Rushforth, a distinguished psychotherapist then in her nineties, how she could be as sure as she was that 'spiritual' experience was a genuine response to a spiritual reality. She looked at me and said simply: 'I have *always* known it.' The experience, when it comes, appears to be compelling.

12

Dealing with Emotions: Some Western Ways

There are three distinct ways in which the human mind can develop. One is by adding a new mode to those already available, as when the line mode begins to appear towards the end of the first year of life. Let us call this 'expansion of the repertoire'. Another is by achieving new kinds of competence within an established mode, as when one learns to cook or to improve one's understanding of the relationship between the earth and the sun.[1] This we may call 'within-mode learning'. The third is by becoming better able to determine the use one makes of one's modal repertoire – to combine modes for a given purpose or to shift from mode to mode at will. The last of these, which we shall call 'control of the repertoire', is the one that has received the least explicit attention, at any rate in Western cultures.

There are certain kinds of voluntary shifting from mode to mode at which some of us are quite skilled. For instance, those who have practised intellectual discipline to any extent have developed some capacity to turn their minds to intellectual activity when this seems desirable. But the general question of how to control the repertoire is not often discussed.

As a preliminary to a consideration of this topic, there are a few things to be said about *involuntary* switching from mode to mode. First, some kinds of moment-to-moment shifting are commonplace – indeed, continually happening – in everyday life. This, I think, is evident. We burn the toast because we start to be concerned with something we plan to do later in the day, or to worry in case we gave someone a 'wrong impression', or whatever.

The point has already been made that the two modes into which most of us slide involuntarily with the greatest ease are the line mode and the core construct mode. They seem to draw us. We flutter and slither towards them. However, it should also be noted that there are times when it is the point mode that suddenly and compellingly takes over.

Just occasionally, this happens in a very agreeable way. Lewis Thomas describes such a happening during a visit to Tucson Zoo. He was walking at the time along a deep path which is cut between two glass-walled ponds so that a family of otters and a family of beavers can be seen very close to one's face – within reach, but for the glass – as they swim below the water and rise to the surface. Thomas says:

I was transfixed. As I now recall it, there was only one sensation in my head: pure elation mixed with amazement at such perfection. Swept off my feet, I floated from one side to the other, swiveling my brain, staring astounded at the beavers, then at the otters ... I remember thinking, with what was left in charge of my consciousness, that I wanted no part of the science of beavers and otters, I wanted never to know how they performed their marvels ... All I asked for was the full hairy complexity, then in front of my eyes, of whole, intact beavers and otters in motion.

It lasted, I regret to say, only a few minutes and then I was back in the late twentieth century, reductionist as ever ... I became a behavioral scientist, an experimental psychologist, an ethologist and in the instant I lost all the wonder and the sense of being overwhelmed. I was flattened.[2]

Thomas saw himself as escaping in this moment not from the line mode but from the intellectual modes habitual to him. I am sure, however, that in his few seconds of exultant point-mode activity, line-mode concerns were also swept away. All he wanted – and briefly got – was full, joyous, undivided involvement in what was directly before him. And when he returned unwillingly to the intellectual modes, the thoughts that filled his mind for a while were not about the animals but about his own response to them. His purpose towards them, he decided, had been simply friendship.

Here is another quite similar example taken from the writings of Marion Milner:

One day I was idly watching some gulls as they soared high overhead. I was not interested, for I recognized them as 'just gulls' and vaguely watched first one then another. Then all at once something seemed to have opened. My idle boredom with the familiar became a deep-breathing peace and delight, and my whole attention was gripped by the pattern and rhythm of their flight, their slow sailing which had become a quiet dance.[3]

Both of these examples happen to be about watching animals, but moments of intensely satisfying immediate experience can, of course, occur in many contexts. However, point-mode dominance is also, and more reliably, produced in all of us by moments of acute crisis. For example, if a plane has crashed, the passengers are concerned to get out of it. And for a while nothing else matters. A passenger who was fretting ten minutes previously about the risk of a crashed marriage or about the recent crash of some speculative investment will leave the line mode abruptly and concentrate entirely on what is here and now: the distance to the nearest exit and the business of getting there.

Some people are capable of great unselfishness in moments of crisis. They may help others to the nearest exit at the risk of never reaching it themselves. They may also have flashes of acute concern about those who will be bereaved if they die. And there are some reports of vivid flashbacks of the course of one's life. Nevertheless, at times of great danger or need most line-mode concerns are immediately superseded. The mind sheds them. The point mode takes over.

For many people in the world today it is fairly uncommon to face immediate crisis. I do not forget the millions for whom external circumstance is perilous or painful in a moment-to-moment way, with little respite. But very frequently it is the activities of the line and core construct modes that bring human suffering. This is not new. The Buddha understood it well.

Awareness of the troubles that these two modes bring should

not obscure the fact that we could not lead human lives without them. We need them both. And it would be a gross distortion to neglect the happiness and enrichment they give us. But they also cause turbulence, conflict and suffering – and we need ways of coping with the pain. The nature of this problem has been considered in earlier chapters but it may be a good idea to restate it now in preparation for the rest of the argument. Briefly, then, we are so made that we generate many diverse purposes; and we may pursue these with intense passion and tenacity. However, the nature of existence in space-time is such that one event is frequently incompatible with another so that, often, one purpose can succeed only at the expense of another's failure. Thus clash and conflict are unavoidable, both within a single mind and between different persons – or social groups – with different goals. As well as all this we have cognitive capacities that give us knowledge of a wide range of hazards. These include not only dangers to the physical self but threat to the self-image that we hold so dear. The self-image is a construct in which, normally, thoughts and emotions are inextricably intertwined. It belongs in the core construct mode, together with similar notions of *our* family, *our* social group, *our* religion, *our* rights – and the wrongs that have been done to us or may be done.

It is obvious that the line and core construct modes often work closely with one another. Together they yield much uncomfortable knowledge, from which we are able to find some protection by the use of devices known as 'defence mechanisms'. These are, in general, ways of deceiving ourselves. They start to operate while we are still very young (probably during the second year of life)[4] and while we are neither aware of what we are doing nor able to foresee the troubles that our primitive defences may later bring.

The upshot of all this is that much human suffering stems not from any present external circumstance but from the ways in which we deal with that which has been and that which may be – or with that which, like death, surely will be.

The question then is: what can we, as adults, do? Is there any help for it, or must we simply accept that such suffering is beyond our power to control? This amounts to asking whether we can learn to improve upon the self-deceiving defences of childhood.

Notice that regulating the emotions can mean much more than getting rid of undesirable ones. It can mean learning how to achieve positively valued ones, like serenity and peace. This is a human dream of great antiquity. Consequently there are various sources of advice, ancient and modern.

I should at once make it clear that, in what follows, I shall not be attempting a review of the range of 'therapies' now on offer. I shall not be talking about disturbances of the kind we call 'mental illness' but rather about the human suffering with which each one of us must somehow deal. Of course, the line separating mental illness from 'normal' trouble is not sharp. It partakes of the general fuzziness of category boundaries. However, I shall intend to stay over towards the normal side.

The first thing to recognize is that those of us who live in twentieth-century Western cultures are heirs to somewhat confused and limited ideas about the possibility of emotional control. On the one hand, the belief that we have some power to regulate what we think and how we act but not how we feel has wide currency. We say things like: 'I'll go and think it over', or: 'I decided to have a think about that'; whereas we do not usually talk about deciding to have an emotion.

I am talking here of popular, largely unexamined notions – of that which is taken for granted by most people, most of the time. More extreme positions have been adopted by some intellectuals. The notion that we have any real power of choice at all in matters of behaviour or experience came under strong attack in the 1870s and 1880s when social Darwinism produced what Hughes calls 'a kind of scientific fatalism'. Carried to its extreme, this led to statements like: 'Virtue and vice are products like sugar and vitriol', or: 'Nature and history are only the unrolling of universal necessity'.[5] Somehow it seemed at the time enlightened to hold such views.

As we saw earlier, similar beliefs came to be dominant in that part of psychology known as behaviourism; and for a while they had a wide influence in the discipline. However, I am sure that few, if any, people were ever *in practice* wholly guided by such ideas. The immediate experience of deciding what we are going to do next is too strong. On the other hand, we seem to have no comparable conviction where emotion is concerned. 'I can't help my feelings', we say, and we are generally not contradicted.

When this remark is made the reference is very often to the emotions of love, hate or anger. However, it is evident that we do also urge one another to attempt certain kinds of control in ways that imply some belief in the possibility. 'Cheer up', we tell people, or: 'Keep your cool'.

There is also a recognition, going back far beyond Freud and strongly expressed in some branches of literature, that emotions can at the very least be denied outlet. For instance, Jane Austen talks repeatedly about the control of emotions by 'exertion'. This is for her a highly conscious and deliberate activity which she evidently admires and values. In *Sense and Sensibility* the contrast between Marianne and Elinor is drawn to a considerable extent in terms of the capacity to 'exert oneself' when passionate distress would otherwise get out of hand. Thus when Elinor tells Marianne that she has known about Edward's secret engagement to Lucy for four months without letting the pain show, the success of the concealment is attributed to 'conscious and painful exertion' and is presented by the author as a triumph. Marianne, who is not at all used to attempting such feats, thinks at first that Elinor cannot really have cared very much, but Elinor assures her that she has suffered most keenly. She was, however, supported by 'feeling that I was doing my duty'. And the reader learns that during the period of exertion Elinor *told herself* many things, such as that 'it is not meant – it is not fit – it is not possible' for one's happiness to depend entirely on any particular person. The outcome in the end is declared to be that her suffering is not merely hidden, it is genuinely overcome: 'I no longer suffer materially myself.'

Later events suggest that this claim may be overstated, but the aim at real mastery is clear, as is the belief in its possibility. What Jane Austen is telling us is that painful emotion can be controlled by the direct efforts of the will. 'I *will* be calm; I *will* be mistress of myself,' says Elinor later, when Edward unexpectedly arrives. In these exertions Elinor's strength seems to derive mainly from the functioning of the core construct mode – from a system of beliefs about decorum and social obligation and from strong attachment to a self-image shaped in accordance with them. She does also use thoughts that go beyond notions of social acceptability: she tells herself that 'it is not meant' and 'it is not possible', as well as being 'not fit'. That is, she talks to herself in ways clearly meant to change the positive evaluation – the supreme importance of Edward in her life – that is the source of pain. And yet it turns out that, even if she subdues her suffering by these means, her love for Edward is in no way diminished. For when she discovers that he is free of Lucy – honourably free, of course – she goes into 'a state of such agitation as made her hardly know where she was'. So what Jane Austen describes for us falls very far short of any successful thoroughgoing emotional regulation. Love survives the exertions; and what is more, in certain circumstances wild joy is fully permissible. Yet decorum still requires that Elinor, even at this peak of happiness, should rush out of the door before she bursts into tears.

Marianne, on the other hand, represents the full flow of romantic sensibility that had been described admiringly in the eighteenth-century 'novel of sentiment' by writers like Lawrence Sterne and Henry Mackenzie. This went so far that it led to a devaluation of emotional control, which was considered not so much difficult as undesirable. Heroes or heroines tended to be warm, impulsive and overwhelmed by their emotions. Villains were characteristically 'calculating', and they were cold deep down, whatever they might contrive to seem. Thus those who hid their emotions were, on the whole, not to be trusted. They were not held up as models, worthy of emulation; and this part of our tradition is still strong in spite of later reactions against it.

However, there is another part of our cultural heritage which, paradoxically, fits ill with the romantic image in spite of the fact that it is often seen as the very height of 'romance'. The orthodox marriage service contains a promise to love until death. Yet if this is not accompanied by a belief that emotion can be controlled, then the promise is worthless. A promise that *cannot* be kept had better not be made. In practice, of course, it is often not so much a promise as an expression of faith that present feelings will endure.

If we turn from the question of love for an individual to consideration of more general attitudes to others, we find that the Christian religion gives us the 'new commandment' that we 'love one another'. Also we are traditionally taught to ask that our own trespasses be forgiven 'as we forgive them that trespass against us'. But true forgiving means ceasing to *feel* resentment or any desire for revenge. So again there is conflict between a powerful part of our cultural heritage – still powerful in spite of waning adherence to the Church – and the prevalent belief that emotions are beyond our control. Clearly some reconciliation is needed.

This is sometimes attempted by taking the traditional injunctions as referring to action, not to feeling (though many Christians would not regard it as legitimate to do so). On this view we are not really being enjoined to love, only to act *as if* we loved; and we are not being told to feel no bitterness, only to act as if we felt none. But then we are still in trouble, for 'acting as if' is what calculating villains do, is it not? Also there is the reasonable doubt whether acting can be sustained for long.

Against these arguments others may be set. First, the motive matters. Calculating villains 'act as if' in order to deceive. But one might try to change one's conduct in the hope that by doing so one would effect deeper change in oneself. Is this naive?

Thomas Nagel says, rightly I believe: 'If we are required to do certain things, then we are required to be the kinds of people who will do these things.'[6] But how are we to become different kinds of people, different not just in how we behave but in the deep structure of our being? Is it absurd to think that acting as if we were already as we want to be might promote the desired change? Bruno Bettelheim, while he was a prisoner in Dachau and Buchenwald, came to think that this is not an absurd notion at all. He says: 'Only dimly at first, but with ever greater clarity, did I also come to see that soon how a man acts can alter what he is. Those who stood up well in the camps became better men, those who acted badly soon became bad men . . .'[7]

A 'better person' has to be one whose emotions are in some sense better. For emotions are our value feelings – our direct responses to the recognition of importance. And personal moral quality cannot be separated from the question of what is held to matter. This is surely involved, even if it is not all that is involved.

So far, then, there have emerged two suggestions about how we might change in order that we can be as we want to be so far as emotion is concerned. (It is a big assumption that we can easily know how we 'really' want to be, but let us for the time being proceed without questioning this.) Both suggestions depend on the belief that emotion can come within the scope of what we ordinarily mean by volition. Jane Austen's heroine simply tries to 'exert' direct willpower; and the process seems rather like what Marion Milner calls 'an internal clenching and grunting';[8] or at least this fits Austen's description of it as 'painful'. On the other hand, Bettelheim's observation of behaviour under extreme duress yields a glimpse of a more indirect route by way of the control of action.

Movement of the 'voluntary musculature' is what we most readily perceive ourselves as being able to control. So if this provided a good way of changing our emotions for the better, it would be attractive, for we would have a starting-point of established competence. It may be as well to begin, then, by considering what people can do already, with little or no trouble at all.

How we come to achieve control of our movements we can none of us remember; nor do we have any direct knowledge of how we now do it. Physiologists can tell us a few things about what goes on, but we do not experience these inner bodily events. If we decide to wiggle our toes or raise an arm, it usually just happens. There is nothing conscious between the intention and its realization. On the other hand, some of us can remember extending the basic control to new muscle groups. I do not refer to extension in the direction of complex skills such as cycling or piano playing, but only to abilities like wiggling the ears instead of the toes. Most people cannot do this, some have learned.

I cannot wiggle my ears but I have a personal memory of learning to raise one eyebrow at a time. When I was perhaps eight years old I saw someone do this and made up my mind that I was going to learn to do it too. Whether I asked anyone for advice I am not now sure, but I think not. What I remember is spending periods of time in front of a mirror just trying.

The interesting question is: what did the 'trying' consist of? I find it impossible to say. I just tried. All I had was a clear goal and one important bit of understanding, namely that a mirror would help to give me knowledge of results – or 'feedback', as we would now say. How I had come to know that feedback matters is unclear. Perhaps we must call this an intuitive insight. At any rate the mirror seemed important for the success of the undertaking – and it did succeed. I became able to move my eyebrows separately, and in rapid rhythmic patterns. It was very satisfying. Without the mirror I would still have had feedback, of the kinaesthetic variety. But the addition of the visual feedback probably gave me evidence of the first signs of success, the first tiny independent movements. At the very least it must have confirmed the kinaesthetic cues.

What I had discovered, in effect, was the principle underlying the techniques now known as biofeedback. Only I did not realize the full scope of my discovery. I did not generalize my knowledge. The generalization had to wait till Neal Miller applied his mind to the subject in the 1960s.

The idea of biofeedback is that much bodily activity which lies beyond our ordinary control can be regulated at will if we get information about each very slight change that occurs during our attempts at learning. And the possibility of regulation has been found to apply not just to uncommon levels of control of groups of voluntary muscles, as in the case of eyebrows or ears, but to the regulation of bodily functions not usually within the scope of our volition at all. Thus people can learn to lower blood pressure, to alter heart-rate and so on, if only they can be given precise information about tiny changes. To achieve this, measuring instruments are used and these are made to operate clear signals. For instance, a musical note may sound to indicate that the heart-rate has slowed down ever so slightly. The learner then experiences the task as that of producing the sound and every bit of success is instantly known. Under these circumstances learning is often fast and efficient.

Now the kinds of physiological function that may be controlled in this way are intimately related to what we experience as emotion. If we are angry or afraid, the heart beats faster, and the rapid, pounding beat is a component of what we feel. It is, however, only a component; for the heart also beats fast if we feel joyful excitement.

In spite of this evident fact, some have claimed that the bodily sensations *are* the emotion, if the full pattern is considered. William James and Carl Lange are generally regarded as the original advocates of this belief. For example, James argued that we feel sorry because we cry instead of crying because we feel sorry. But of course, that leaves us with the question: why do we cry?

James was well aware of this and he did not overlook the role of some perceived 'arousing event' in the production of tears. What he really wanted to stress was that 'without the bodily states following on the perception the latter would be purely cognitive in form, pale, colourless, destitute of emotional warmth'.[9]

He was quite right in this as far as it went: emotion is of the body. Emotions are value *feelings*; and there also exist value judgements which by themselves are relatively cool and pale. However, James did not give enough weight to the interpretive or evaluative component of emotional experience; and much subsequent research has shown this to be so.

But let us get back now to the topic of control. If we can learn by biofeedback techniques, or in any other way, to moderate at will certain bodily sensations like heart-beat, how much *emotional* control do we gain? If the heart stops racing or palpitating and becomes steady and slow, what then? Shall we still feel angry, or afraid, or excited and elated, as the case may be?

The answer is that the slowing of the heart will affect the emotional experience. Something of intensity or urgency will leave it. But, if nothing else has changed, we shall still be left with the positive or negative assessment of importance and this will continue to be 'coloured' by some feeling: a feeling of annoyance perhaps, or calmer pleasure.

The conclusion must be that anyone who learns to slow the heart at will, or to regulate in a similar way any other physiological indices of arousal, has indeed achieved a measure of deliberate emotional control. So this is a resource that is available in principle to anyone. However, it seems, at least on first reflection, to be quite limited in scope. It can 'take the edge off' certain acutely unpleasant experiences. But can it do more? Can it contribute in any way to the establishment of positively desirable ones?

This is an important question that will be raised again later. Meanwhile there is another relevant competence over which we all feel we already have a measure of voluntary control, and that is speech. We can to a considerable extent decide what we are going to say and say it. This, of course, entails use of the voluntary muscles. It may also enter into the business of 'acting as if', which has already been mentioned. For what we say is evidently part of how we act.

Control of overt speech is by no means perfect, even apart from such problems as stammering, or the various kinds of aphasia. Nevertheless, most of us most of the time choose our words quite well. And, perhaps more importantly for the present topic, we can regulate our inner speech. We have some control over what we say to ourselves 'under the breath' or in total silence.

We do habitually use inner speech − or occasionally outer speech − to give ourselves certain kinds of command or instruction. Recall how Elinor says (inwardly, of course): 'I *will* be calm. I *will* be mistress of myself.' Elinor is a model of sophisticated striving for one kind of self-control. But I also recall at another extreme an occasion when a four-year-old girl called Anna was sitting beside me on a chair too high for her, which left her feet dangling. Anna kept wriggling about and her feet kept knocking against one another. 'Stop fighting, feet!' she suddenly said, looking down with a stern expression on her face. The 'fighting' was in fact, I think, a bodily expression of some emotional unease that Anna was feeling at the time. It was perhaps for this reason that words were called in to support the kind of instant decision that usually controls the skeletal muscles. But the words were not very effective. The feet stopped obediently for a moment or two then kicked one another again as before.

The truth is that direct verbal commands are usually quite limited in what they achieve when it comes to matters emotional, whether the commands come from someone else or are self-given. We tell ourselves to cheer up and the gloom remains or soon returns. We tell ourselves not to be afraid and fear does not go away. If, like Anna, we command our *bodies* instead, we may fare rather better. The instruction 'Smile!' can usually be obeyed; and there is some evidence to suggest that if we manage to assume a happy expression and keep this up, we may come to feel happier. So Bettelheim's claim about the effects of 'acting as if' arises again.

Nevertheless, we often do not succeed in changing our emotions by direct command. And there is a puzzle here. Why should this be, when we are so successful in some ways at manipulating our own awareness, at persuading ourselves, at choosing what we want to know and experience? Why do we sometimes seem to respond so wholeheartedly to the inner voice and sometimes barely at all?

The answer to this question must be complex. Many circumstances will affect the outcome. However, there is a distinction that turns out to be crucial: the distinction between telling ourselves that something is the case and telling ourselves *to do* – or *not to do* – something, where the doing entails change in some emotional state. It is the latter kind of telling that often seems to produce no effect at all or even an effect opposite to the one intended.

The reason is not hard to find once the distinction is drawn. The sphere in which verbal command is most generally efficacious is that of the voluntary musculature. It is true that we do not *need* words for the control of our own muscles, and also that sometimes, as in the case of Anna's feet, self-directed verbal commands even in this sphere are not thoroughly effective. We do, however, regularly use verbal commands in our attempts to regulate the musculature of other people, and I think there is no doubt that when we tell *ourselves* to do things, the system that tends to come into play is the system of voluntary muscle control. Yet this system proves to be inappropriate for dealing with our emotions. We will the effect – and nothing happens. Indeed, the harder we try *in this way* to dismiss an unwanted feeling, the more conscious of it we are likely to become. The experience is disconcerting.

Consider, on the other hand, the effect of telling yourself that something is the case – for example, of saying: 'I feel calm and confident', although you do not. Certainly this does not always lead to an increase in calmness and confidence. Sometimes another inner voice promptly says: 'But I don't'. Also external circumstances, like the presence of a tiger ten yards away, may well count for more than any efforts at self-persuasion. Nevertheless, it appears to be a general truth that we are disposed to believe ourselves when we make assertions as to the way things are. Although this disposition can be outweighed, it is powerful; and in certain circumstances it can become extremely strong.

Our tendency to this kind of belief is put to use in many techniques for achieving relaxation. It is well known that the repetition of: 'My right leg feels heavy and warm' tends to induce in the right leg the feelings described. In favourable circumstance the leg *becomes* heavy and warm. The muscles relax, the blood flows more freely. The body starts to conform to the description given.

In trying to understand this it is instructive to consider that, when children start to use language, they first meet with such descriptions as accounts of actual feelings. That is, the feelings come first, then the language, once learned, is used to talk about them. The feelings initially give rise to the descriptions. What is remarkable is that, later, the possibility arises of movement in the opposite direction: the descriptions become able to give rise to the feelings. If we make use of this possibility, then, in addition to telling ourselves what we feel, we can to some extent at least feel what we tell ourselves.

This is the principle that underlies a whole range of techniques known as 'autogenic training' or 'self-hypnosis', or just 'relaxation'. These techniques often make much use of imagery, just as hypnotists commonly do. Words are then used to evoke the images. Thus instead of saying: 'My leg feels heavy', one might say: 'My leg is made of lead'. So these ways of achieving a calm, relaxed body shade over into what is sometimes known as 'creative visualization', where much more of the emphasis is on the voluntary production of imagery. This may still be imagery relating to the body, as when people with cancer are encouraged to think of their immune systems as armies of fierce creatures – sharks, perhaps, swimming in the blood – that can eat up the tumour. It is claimed that the more vivid and powerful the imagery the greater the chance of a successful fight against the disease. Here again the principle being invoked is that the body somehow conforms to our description of it, whether that description is made of words or of 'pictures in the mind'.

Sometimes, however, creative visualization is not concerned with the direct imagining of a bodily state. It may rather be that an image is constructed which has the desired kind of association. As well as memory images, we all have images that we create or invent; but the distinction is not as sharp as it might seem. Our memory images are in general not like photographs, they are constructed. On the other hand, invented images always have some basis in things experienced, however striking their novelty. Even when we create by negating what we know or by standing it on its head, we do not create *de novo*. Thus images are powerfully linked to the kinds of circumstance from which they originally stem.[10] It follows that the image has some power to bring with it the typical experienced context: if we imagine a still lake at dawn, a faint mist on the water, the sky glowing in the east and sounds of birdsong, this is likely to induce a calm and pleasant emotional state (unless we have some personal memory linked to an emotion of a different kind). Calling up a suitable image and dwelling on it is thus a means of emotional control. By the same token, if powerful images arise unbidden, then emotional control is lost. The imagery of our dreams normally eludes our conscious control though in one sense we produce it: and it can give rise to strong emotional experience, including extremes of terror. Also memory images can surge up in us, triggered perhaps by a chance event.

These last two topics – control of what we tell ourselves and control of imagery – are related to the question of the control of thought. For both words and images are used in thinking.[11] Earlier in the chapter, speaking of popular conceptions, I said that people in our culture seem to believe that thoughts are controllable – or at least more so than emotions.

We have to recognize that the control of thought is itself far from perfect.[12] However, it can be considerable. And the question now is whether our thoughts are promising as a means of controlling our emotions. Are they an important resource available to us for that purpose if we know how to use them well? There is a children's story – a kind of modern fairy-tale – about a monster called the Ugsome Thing who had the magic power of enslaving anyone whom he could cause to become very angry. Consequently he soon had all the servants he needed, except that he lacked a good washerwoman. One day he noticed a cottage garden full of clothes that were whiter than white, and he thought the old woman who lived there would do very well. So he cut her clothes line and waited for her fury. But when the old woman saw what had happened she just said quietly that it was fortunate the clothes line had broken on that particular day, for her chimney had been smoking and there was sure to be soot on the clothes, which meant she would have had to wash them again anyway. She then returned to the wash-house 'singing as she went'. After this the Ugsome Thing tried more and more extreme measures, culminating in the burning of her cottage. But each time the woman found some way of regarding the mishap as fortunate. 'How lucky I am,' she said again and again. She had a habitual way of thinking – and of talking to herself – that kept her constantly serene.[13]

This is the kind of outcome aimed for by the techniques of

cognitive therapy, though the goals are generally more modest than cheerfulness in the face of any disaster, however great. Cognitive therapy and its variants, such as rational-emotive therapy, are attempts to help people to think differently about the situations they find themselves in, and so to feel differently about them. In the course of changing how they think about things, people are encouraged to talk to themselves in new, more positive, constructive ways.[14]

It is important to notice how this differs from the kind of 'telling oneself that . . .' which was discussed earlier. The distinction depends on the fact that emotions generally have two components: an evaluative component and a feeling component (though these are not normally experienced sequentially, but as one 'value feeling'). If we tell ourselves that our muscles are relaxing, that our heart is beating quietly and steadily, that we are breathing peacefully, then we are affecting emotion – if the attempt works – by way of action upon the bodily feelings. If we tell ourselves that we are fortunate in regard to some event – or if we try in other ways to achieve more optimistic and fewer gloomy judgements – then we are affecting emotion by way of the process of evaluation. We are taking a different view of what matters.

Either way can be tried by anyone. And they are in no way mutually exclusive.

The ways so far considered of trying to bring our emotions into order vary in a quality I have been calling 'directness'. Thus it is arguably more direct just to *will* change than to try to achieve change by learning biofeedback techniques or by deliberate imagining or talking to yourself, or by acting as if you feel what you do not feel in the hope of eventually coming to feel differently. However, all of these methods still share a certain directness, at least to the extent that they identify something wrong now ('I don't like or approve of the way I feel') and try to tackle the problem without worrying much about why it arose in the first place. I do not mean that they ignore causes; but they tend to deal in immediate causes – tension in the muscles, perhaps, or bad ways of breathing or habitual pessimistic notions. Typically there is no attempt, or not much attempt, to look back into the past, asking when and why the trouble began. And certainly there is no assumption that it began in early childhood. In contrast psychoanalysis, that great twentieth-century attempt to redeem the line mode, has been called 'archaeological' because it entails a 'dig'. This dig, which is often prolonged, has the aim of recovering what may reasonably be considered as ancient 'artefacts', for they have been constructed by the child in the course of early dealings with the world and then overlaid – that is, lost to consciousness. Typically such artefacts are held to be the products of the defence mechanisms: they represent what the child did when forced to face the fact that reality does not afford all that is desired.

Freud evolved a detailed developmental theory about the nature of the child's early desires and of the developing conflicts. This theory is crucial for an understanding of his own therapeutic practices and of the practices of those who, coming after him, called themselves 'Freudians'. However, there are general characteristics of the psychoanalytic procedure that do not depend on whether Freud was right about, say, the instinctual sexual-aggressive nature of the child's problems or the value of dividing the 'psychical apparatus' into id, ego and super-ego. Central among the general features is the belief that recovery of an artefact represents a triumph: that which was hidden has been brought to light. Thus it may now be examined and dealt with by the mature mind.

This seems a very reasonable claim if one accepts the basic idea of the defence mechanisms; and it has proved widely appealing. Great hopes have been based on it. Vast time and effort have been expanded on it, for the digs tend to be lengthy. But have they proved to be worthwhile? Even Freud expressed doubts about this towards the end of his life. I quote from *Analysis Terminable and Interminable*, a particularly interesting late reflection on his work:

. . . it is always a question of the quantitative factor, which is so easily overlooked. If this is the correct answer to our question, we may say that analysis, in claiming to cure neurosis by ensuring control over instinct, is always right in theory but not always right in practice. And this is because it does not always succeed in ensuring to a sufficient degree the foundations on which a control of instinct is based. The cause of such a partial failure is easily discovered ... If the strength of the instinct is excessive, the mature ego, supported by analysis, fails in its task, just as the helpless ego failed formerly ... There is nothing surprising in this, since the power of the instrument with which analysis operates is not unlimited but restricted, and the final upshot always depends on the relative strength of the psychical agencies which are struggling with one another.[15]

So Freud clearly acknowledges the limits of his methods. Life, in his view, is an unending battle against instinctual urges, and we can never be certain that they will not prove too strong. Thus the return to consciousness of that which had been hidden is no guarantee of success.

Bettelheim's observations of behaviour under extreme conditions in Dachau and Buchenwald have already been mentioned.[16] Growing up in Vienna between the wars, he had been much impressed by what psychoanalysis seemed to offer. He had himself spent years being analysed. He was then shocked by the discovery that in the camps psychoanalysis proved not to work as expected. He decided that it was 'by no means the most effective way to *change* personality', adding that while psychoanalysis 'told much about the "hidden" in man, it told much less about the "true" man'.

Bettelheim does, however, record with gratitude that psycho-analysis as a method of observation, rather than a body of theory or a therapy, helped him to understand what might be going on in the unconscious minds of prisoners and guards – 'an under-standing that on occasion may have saved my life, and on other occasions let me be of help to some of my fellow-prisoners, where it counted'. He concluded that the trouble with psychoanalysis lies in its concentration on what has gone wrong. In particular it has little to say about 'what makes for "goodness" or "greatness"' in a life. This, I think, is true. And the trouble is inherent. It consists in the fact that the process of analysis – the archaeological 'dig' – encourages, even requires, so great a concentration on the line and core construct mode.

Recall the three kinds of development: expansion of the repertoire; within-mode learning; and control of the repertoire – that is, development of the ability to move flexibly between modes at will. Prolonged analysis of the events of one's own past life and their significance is not in itself conducive to the cultivation of other modes, nor yet to the growth of that most important and much neglected capacity for deciding, at any given time, in what mode one wants to be.

This is not to deny the value of reflecting on how one came to be as one is and trying to be free of old self-deceptions. The hard acknowing that this entails – the kind of established honesty with the self that may result from it, if it succeeds – are no mean achievements. They may indeed be prerequisite for certain sorts of further advance. But there are dangers. 'Interminable analysis' may mean endless absorption in one's own past of a kind that narrows the vision and is not morally desirable. Also, as Freud finally acknowledged and as Bettelheim was forced to see, analysis does not guarantee that, even for the self, all manner of thing will thenceforth be well.

Julian of Norwich, who had so firm a belief that all would be well, urged that when we suffer pain we should not 'pursue it in sorrow and mourning for it' but rather 'suddenly we should pass it over'.[17] That is, we should do our best to lay it down and move on.

13

Dealing with Emotions: Some Buddhist Ways

The last chapter was about some of the ways in which we may be able to reduce the suffering that so often arises from the functioning of the line and core construct modes: from either or, more usually, from both working together. The possibilities so far considered have been, for the most part, attempts to achieve improvements *within* these modes, to bring them somehow into better order or greater maturity. Thus, for example, Freudian psychoanalysis aims to redeem both modes by a kind of reconstruction, following upon insight into what originally went wrong. And cognitive therapy aims at reconstruction without bothering so much about developmental history. However, there remains another kind of possibility: instead of trying to redeem a troublesome mode we may try to leave it, at least for a while. We may concentrate on getting out. The range and efficacy of the options for doing this will depend on what other modes we have available and also on the state of our capacity for control of the repertoire – or between-mode switching – which, like most human skills, may develop or may remain relatively immature.

All of us have at least some variety of the point mode to escape into. So it comes as no surprise that, when children first begin to consider it possible to take any action against distressing emotion, their proposed solution is to *do* something practical: to get a favourite toy, to go out to play. To the extent that they can then *keep their minds* on the toy or the game they will be functioning in the point mode.[1] Paul Harris, reporting research on this subject, says that some six-year-olds (the youngest group studied) were quite pessimistic about the possibility of helping themselves towards a better emotional state; but those who thought they could do so generally spoke of finding some new activity to engage in. Older children also tended to suggest 'changing the situation', as Harris puts it. But they were more often able to explain how this would help: it would enable them to forget what was upsetting them.

What this amounts to is that the children were describing a switch from the line mode into the point mode. And some six-year-olds already had the same strategy available although they could not give any account of its value.

In our culture, at least, the resort to point-mode activity remains at all ages the commonest means of escape. Occupational therapy is an organized example of it. 'Get something to take your mind up' is frequently given advice, and it generally means: do something actively physical, grow cabbages or play golf. Kipling's advice, given to 'kiddies and grown-ups too' was

> . . . not to sit still,
> Or frowst with a book by the fire;
> But to take a large hoe and a shovel also,
> And dig till you gently perspire . . .[2]

However, for some people the intellectual modes also provide a way out. Whitehead has called mathematics 'a refuge from the goading urgency of contingent happenings'.[3] A number of boys who took part in research done by Harris and Guz at English boarding schools suggested that 'getting stuck into work' or 'putting your mind on the prep' (i.e. on study) might help them in dealing with the deep distress and longings for home that many felt. One said: 'When I felt a bit homesick I started doing mathematical puzzles.'[4] Nevertheless, it was also freely acknowledged by these children that the strategy of becoming absorbed in something else – whether games or study – was not one that they had fully mastered. Sometimes, they said, you couldn't concentrate because of feeling unhappy. And, worst of all, when night came and you were necessarily silent and inactive, the pain would surely return.

Harris comments on the way such boarding schools pack the daytime hours with organized and very public activities. There is no encouragement of solitude and silence. And by the same token no education is given in how to use these well.

'Not to sit still,' says Kipling, a product of this system. But there is a whole different human tradition in which *sitting still* is held to be an essential feature of the effort to control suffering and thus to be a way in which many hours of a life can properly and profitably be spent.

Before we go on to consider this alternative idea I want to raise a new question: if we are going to put any effort into the deliberate development, by whatever means, of the ability to switch from mode to mode at will, is the aim merely to reduce our own distress and increase our own happiness? Or are there other reasons why such an enterprise might be judged worthwhile?

There is, first, an obvious distinction to be drawn between aiming to reduce one's own unhappiness and aiming to find means of reducing the suffering of all humanity. The children that Harris describes were concerned in modest ways with the former. The Buddha was concerned in ambitious and far-reaching ways with the latter.

It is easy to see that these two enterprises are vastly different. But another distinction now becomes relevant. This is the distinction between the aim of finding ways to reduce suffering, either for oneself or for others, and the aim of self-development or self-transformation. In other words, are we talking only about becoming happier, or also about becoming in some sense wiser or *better*? Are we talking about such things as liberation or enlightenment?

What is of great interest is that an initial concern for the reduction of suffering seems often to lead people well beyond this relatively modest goal. There tends to develop, if the effort is serious and sustained, an exploration of the possibilities for human experience that can lead far from the starting-point. This is a matter of very great importance often not understood today.

At this point in the argument I propose to spend some time on a discussion of Buddhism because it is one of the most sustained and widespread enterprises of this kind about which we have any knowledge. Buddhism is a particularly resolute attempt to develop skill in leaving undesirable modes at will. Buddhist teaching does not merely say that this development is important, it gives much explicit guidance about how it may be cultivated.

I must at once acknowledge that Buddhism is a vast topic and one where I hesitate for lack of expertise. Yet I think the argument of this book requires its consideration. I can make no attempt at thoroughness and much will be left out. The best I can do is to try not to misrepresent what goes in, though even this is notoriously hard for one brought up in a very different tradition.

The Buddha is said to have grown up in a privileged family, enjoying much comfort and security but feeling deeply troubled by the misery that he saw around him. At the age of twenty-nine he appears to have left his wife and young son and become a wandering monk in an effort to gain understanding. Tradition has it that, after six years of vain effort, he sat down one evening beneath a tree and vowed not to move until enlightenment came and the causes of the wretchedness were clear to him. When dawn arrived, he had reached a perfectly clear understanding of the sources of suffering and the means of release from it.[5]

This was the start of the vast proliferation of human effort and commitment that constitutes Buddhism in all its diversity. Because of this diversity there is much that is disputed, even about the nature of the Buddha's basic experience, and about how it was – or ever could be – described. Zen Buddhists, for example, consider that words cannot capture any experience adequately, let alone this one. So they have claimed that the Buddha communicated his great discovery to his disciple, Mahakasyapa, by holding up a flower and saying nothing.[6] Earlier doctrines, however, incorporated in the Indian Pali canon, are based upon the belief that the Buddha went to the deer park just outside Benares and there made very explicit statements about what is wrong with the human condition and what is to be done. These have come to be known as the Four Noble Truths. The central notion that emerges is that suffering comes from a concentration on our own desires and from a failure to understand the futility of pursuing them.[7] The futility stems from the fact that this life within space-time affords us nothing that is permanent. Pleasures can never be grasped and held. Indeed, there exists no enduring self to do the holding. Thus the ego is effectively deconstructed.

Watts gives the following account of Buddhist ideas about what we ordinarily regard as selfhood:

For the ego exists in an abstract sense alone, being an abstraction from memory, somewhat like the illusory circle of fire made by a whirling torch. We can, for example, imagine the path of a bird through the sky as a distinct line which it has taken . . . In concrete reality the bird left no line and, similarly, the past from which our ego is abstracted has entirely disappeared.[8]

This is as thorough a downgrading of the line mode – and of the self-image that we construct through the functioning of the core construct mode – as could well be imagined. It amounts to saying that these modes are the source of all our troubles. They feed us illusions. They cause us to attempt the impossible and to pay for this dearly. So what are we to do?

The Buddhist answer is clear: we are to get out of them. But then other questions inevitably follow. Out of them into what? And how? Two possible escape routes are proposed in the early Buddhist texts and accepted in much Buddhist orthodoxy. Paul Griffiths calls them 'the cultivation of tranquillity' and 'the cultivation of insight', the first being an attempt to overcome *attachment*, the second an attempt to overcome *ignorance*.

The notions of ignorance and attachment are both central to Buddhist theory. Indeed, they are almost inseparably linked within it, as we have seen; for the claim is that our ignorance leads us to pursue the ridiculous desires which constitute attachment. And yet it is evident that emphasis on one or other of these notions might suggest different remedies. If you emphasize ignorance, the next reasonable step seems to be the pursuit of better understanding;

whereas if you emphasize attachment, you might be led to concentrate on the giving up of inappropriate attempts to grasp and hold, with all the turbulent emotions that these are bound to bring, given the hopelessness of the endeavour.

Griffiths expresses it thus:

Put very crudely, the cultivation of tranquillity centres upon manipulation of the practitioner's emotional attitudes and the cultivation of insight centres upon manipulation of the practitioner's cognitive skills. The adherent of the former destroys passions by withdrawing from all contact with the external universe, whereas the practitioner of the latter asserts control over the universe by learning to know it as it is.[9]

Thus it might seem that the two routes lead into the emotional modes and the intellectual modes respectively. But this notion, tempting as it initially appears, turns out on closer scrutiny to be a false impression. One is soon forced to recognize that the kinds of cognitive skill which are to be manipulated are not those that modern Western people think of when they hear the word 'intellectual'. 'Learning to know the universe as it is' suggests to us the activities of modern science. Nothing of the kind was aimed at through the cultivation of insight.

There seem to be several crucial points of divergence. First, for the Buddhist the centre of interest lies in human experience – or at any rate this is the starting-point from which and through which the nature of all things is to be apprehended. But the activity of studying this experience is in no way like modern psychology.

King expresses the Buddhist search for insight thus: 'To "see things as they truly are" means, most importantly if not exclusively, to see myself-as-experiencing as I truly am – impermanent, full of pain and impersonal.' So the focus of attention falls upon 'the observance of the body-mind as it functions in its environment'.[10]

But there now emerges a second, and profound, difference between this activity and modern science: in the Buddhist effort after insight, the general nature of the truths to be discerned is assumed at the outset. The way-I-truly-am is taken to be as

declared by the Buddha after his enlightenment. His statements are
contained in the canonical literature, so his discoveries, *considered as
verbal propositions*, are available to all. What, then, remains to be
found out?

It is, of course, the case that all inquiry starts from certain
assumptions, science as we know it being no exception. Nevertheless, science is a method that yields new propositional knowledge.
It can always, in principle, overturn its initial assumptions and
sometimes it actually does so. It is, however, not expected that the
cultivation of Buddhist insight will lead to any such outcome.

It follows that gaining insight is not equivalent to discovering
something radically new. It is not, for instance, like coming to
understand for the first time the double helix structure of DNA.
But if not this, then what? What is it that is to be *cultivated*?

The point is that purely intellectual knowledge of the Buddha's
discoveries is not the goal. To be able to recite the Four Noble
Truths is worth very little, if one stops there. What is to be
cultivated is something based on them but more like 'affective
knowledge' or 'experiential knowledge' or 'intuition' (concepts
not absent from the European tradition, as Fromm points out
when he discusses Zen Buddhism and psychoanalysis).[11]

Yet it would be wrong to suppose that what the Buddha said is
to be accepted unquestioningly even as a basis. Saddhatissa points
out that the Buddha urged his followers to accept his words only
after testing their value for themselves. The idea is that one should
come, by one's own effort, to know the Four Noble Truths 'in
one's bones'. Or, to use a different image – the orthodox one –
they must become like a source of light, illuminating all one does,
one's whole experience. When this happens, one is 'enlightened'.

King speaks of the 'especially Buddhist type of effort aimed at
seeing everything in the light of the Buddhist view that embodied
existence is impermanent, impersonal and unsatisfactory'. And he
goes on to say that 'the experiential realization of this view, when
it is so fully and integrally realized that it becomes the meditator's
existential awareness of his own personal being, here and now,
is called "understanding"'.[12] This understanding is held to be

such that it 'liberates'. It is sought because it offers freedom from bondage.

An important contrast now emerges. It is open to intellectuals who are not on such a path to claim that they want *only* understanding in that they act without the desire to change anything. It is true that they will unavoidably change themselves in some respects if they understand more in any sense; and unless they refuse to communicate their new understanding, which is very rare, they will unavoidably change the minds of others who learn from them. Also some of those others will almost certainly want to apply the new knowledge to control something or other. And yet the originating purpose may be nothing but enhanced understanding. This cannot be the case, however, if the true goal is liberation or enlightenment and if the role of insight is to serve that end. Thus there is a deep paradox inherent in the desire to escape from desire: to the extent that you desire to escape, you are still desiring.[13]

We must next ask what is entailed by the path described as 'the cultivation of tranquillity' and how it fares with respect to this problem. Griffiths gives an account of the four '*jhāna* of form' – or meditative states – by which, according to the orthodoxy of Theravādin Buddhist practice, this cultivation is to proceed. In the first state thus achieved the capacity for reasoning and deliberation remains, together with joy and happiness; but 'desire' and 'negative states of mind' have vanished.

For many of us this would seem a great achievement and a highly satisfactory outcome. But it is only a beginning. In the next state reasoning and deliberation are suppressed, and there remain joy, happiness, inner tranquillity and 'one-pointedness of mind' (that is, a focused condition resistant to distraction, free from conflict or confusion). In the third state joy disappears, to be replaced by equanimity, 'physical happiness' and 'mindfulness'. In the fourth state all happiness vanishes and there remains only 'that purity of mindfulness which is equanimity'.

The process is one of progressive curtailment of experience. Happiness is a staging-post, not a satisfactory outcome. But even

the fourth state was not the ultimate achievement for the schools
of Buddhist doctrine and practice with which Griffiths is
particularly concerned. Indeed, the four *jhāna* of form, though
these are traditionally regarded as characterizing the Buddha's
own way to enlightenment,[14] came to be seen as preliminaries –
first steps on the way to a final state known as 'cessation' which, at
least in certain traditions, is defined as a condition where *no* mental
events occur at all. This may seem rather like death. Griffiths
suggests it was like deep hibernation.

The problem that interests him is how it was then deemed
possible to emerge from cessation, if that condition was ever
indeed reached. For, as he points out, it is hard to avoid the
conclusion that some act of mind would be needed to bring the
practitioner out. This question, though philosophically interesting,
is not relevant here. The bearing of 'cessation' on the present
argument is that it shows how far attempts to escape line-mode
suffering have sometimes gone. The cultivation of tranquillity
may turn into the cultivation of 'mindlessness'. If this succeeds,
there has certainly been escape from desire, but at a price that
seems a very heavy one to pay.[15]

When the cultivation of tranquillity is pushed thus far Griffiths
sees an incompatibility between the two ways because, he says, the
absence of *any* experience has to imply the absence of insight. I
think he overstates the problem. For no one, however adept, was
meant to stay in the condition of 'cessation' for more than a
limited time. The generally accepted view seems to have been –
and still to be – that the achievement of having learned thoroughly
how to cultivate tranquillity (that is, get rid of attachment) is a
help when it comes to the cultivation of Buddhist understanding,
but that the latter – the business of dealing radically with ignorance
and its effects – is the ultimate goal, the one without which the
enterprise is not genuinely Buddhist.

Let us take stock of the argument so far. We have considered
various examples of attempts to deal with unsatisfactory states by
control of the repertoire, that is, by switching from one mode to

another: first, the fairly typical Western method of engaging in some other activity – usually a point-mode one with a significant physical component, occasionally an intellectual one like mathematics; then, by way of sharp contrast, some traditional Buddhist ways of escape.

There is another kind of Buddhism, however, that differs considerably in its recommendations – one in which the aim at 'cessation' has been regarded as a great error and called 'the dark abyss of ignorance'. I refer to Mahayana Buddhism, and in particular Zen, much developed and practised in Japan, though it came by way of China from an original source in India.[16]

Zen Buddhism has intrigued many Westerners, with its baffling 'koans' and paradoxical assertions. Its most striking feature appears at first sight to be its anti-intellectualism. The koan is precisely a problem that the intellect cannot solve. The most frequently cited examples are: 'Let me hear the sound of one hand clapping', and 'Show me your original face you had before you were born'. Students of Rinzai Zen are required by their masters to try to 'solve' these koans – and many more in graduated series if they persist to the point of becoming masters, or Roshis, themselves. But all rational efforts to deal with a koan are bound to fail – and this is precisely the point. Students are required by their masters to keep on trying until they feel totally trapped and despairing and reach what Suzuki calls 'a state of complete standstill'. The hope is that this will lead to some kind of breakthrough or upheaval in which 'thinking with the head' will quite suddenly give way to 'thinking with the belly'.[17]

Suzuki is at pains to stress that this is symbolic talk. What it means is that the koan is to be taken down into 'the unconscious' – or that it is to sink into the whole being. Even this, however, is not the end. The ultimate aim is a further breakthrough in which one passes beyond the 'darkness of the unconscious' and 'sees all things as one sees one's face in a brightly shining mirror'.

This is obviously still symbolic talk. But Suzuki goes on to explain further: the idea is to escape from the subject/object dichotomy that is so strongly experienced when the ordinary self-conscious self knows, feels and perceives, and thus to gain a 'totalistic intuition of the infinite'. This happens 'when the finite ego, breaking its hard crust, refers itself to the infinite'. It is like an intuition of 'something that transcends all our particularized, specified experiences'. The path to this is *through* the deeper layers of the self towards transcendent wisdom. But the intellect is held to block the way – and the block must be removed. The koan is supposed to serve as a kind of dynamite – though it has a very long fuse and the explosion may take months, or even years, to occur.

All of this is apt to sound unintelligible to Western ears (probably to many Eastern ones as well). It is expressed rather more clearly by the modern Zen master Sokei-an Sasaki:

One day I wiped out all the notions from my mind. I gave up all desire. I discarded all the words with which I thought and stayed in quietude. I felt a little queer – as if I were being carried into something, or as if I were touching some power unknown to me . . . and Ztt! I entered. I lost the boundary of my physical body. I had my skin, of course, but I felt I was standing in the centre of the cosmos. I spoke, but my words had lost their meaning. I saw people coming towards me, but all were the same man. All were myself! I had never known this world.[18]

Or as Watts puts it:

When we are no longer identified with the idea of ourselves, the entire relationship between subject and object, knower and known, undergoes a sudden and revolutionary change . . . The knower no longer feels himself to be independent of the known; the experiencer no longer feels himself to stand apart from the experience. Consequently the whole notion of getting something 'out' of life, of seeking something 'from' experience, becomes absurd.[19]

This 'sudden and revolutionary change' is called in Zen terminology 'satori'. It is, I think, equivalent to what the Theravādin tradition calls 'Path consciousness'.[20] Suzuki says that Europeans have no experience 'specifically equivalent' to it. But we shall see in the next chapter that this does not seem to be the case.

We must now note another aspect of Zen that seems to stand in some contrast to all talk of breaking through to intuitions of infinity, though the conflict turns out to be only apparent. This is the emphasis on the desirability of simple, spontaneous, unaffected behaviour. The person who has achieved this Zen goal has 'no conventionality, no conformity, no inhibitory motivation . . . This man's "Good morning" has no human element of any kind of vested interest. He is addressed and he responds. He feels hungry and he eats.'[21]

'When I am hungry I eat; when I am tired I sleep' is a famous encapsulation of the Zen way of living. Action arising naturally, without the intervention of thought and planning, and above all without self-consciousness, is the aim. There is to be a belief – a trust – in the essential 'rightness' of the mind – the original or spontaneous mind, sometimes referred to as the 'Unborn'. This, Watts tells us, is 'the mind which does not arise or appear in the realm of symbolic knowledge'.[22]

From this point of view, then, it is the core construct mode that is seen as the source of great threat. The notion is that the self-image can become totally dominant and overwhelming in one's life so that every action one takes is controlled and contrived, assessed before and after for its implications and consequences. The result can be vacillation and near-paralysis. Perhaps this was Hamlet's problem.

The Zen escape route from this is into the point mode – into the immediacy of the here and now. Eat when you're hungry, sleep when you're tired. Or again:

> Sitting quietly, doing nothing,
> Spring comes and the grass grows by itself.

As Watts puts it, Zen 'does not confuse spirituality with thinking about God while one is peeling potatoes. Zen spirituality is just to peel the potatoes.' The idea is to be fully aware, and absorbed in what one is doing *now*.

The long discipline of the koan can be regarded as one way towards this – a roundabout way, perhaps. Watts likens it to going about for a while with lead in your shoes so as to feel exhilaratingly light-footed when you take them off.

The koan technique is advocated by the school that stems from the thinking of Rinzai. On the other hand, Soto Zen, based on the teachings of Dogen, has a much more direct approach: just stop striving, 'think only of this day and hour'. You should, likewise, 'forget about the good and bad of your nature, the strength and weakness of your power'.

So we have come, it may seem, full circle. We started with the characteristic Western way of escape into the point mode and now we find in Buddhism what looks like the same kind of thing. The Zen solution is *just to peel the potatoes*, which appears at first sight not so very different from taking Kipling's advice and digging in the garden. However, Kipling is proposing a temporary means of escape, while Zen aims at a permanent solution which is, in fact, radically different in kind.

Recall that it is not just the Zen school of Buddhism that recommends the point mode; for we saw earlier that the pursuit of insight in traditional Theravādin practice entails observation of one's own experience *here and now* as it happens. Indeed, it is evident that the Buddhist insistence on impermanence almost necessarily thrusts one in the direction of taking 'here and now' as a primary locus of concern. And this emphasis, always strong, seems to have become very prominent in the twentieth-century development of the Theravāda tradition in Burma where, King tells us, the method of 'Bare Attention' is practised. This method claims descent from ancient Buddhist scriptures on 'mindfulness' but is at the same time quite typical of current trends.

'"Bare Attention" is the clear and single-minded awareness of what actually happens *to* us and *in* us at the successive moments of perception.'[23] It is 'bare' because it simply observes, without reacting 'by deed, speech or mental comment'. If comments arise in the mind, they themselves are to be merely noted and allowed to pass away.

Now, this is assuredly a kind of point-mode activity. But equally clearly it is not the kind recommended by Kipling and resorted to by boys in English boarding schools when the going is hard. So what kind is it? It turns out to be much more like the special kind of objective, detached perceiving that has a crucial part to play in modern science. Indeed, considered simply as an activity it is quite similar.

This may seem surprising. It certainly surprised me when I first realized it. However, the resemblance cannot be doubted.[24] On the other hand, the two kinds of detached observing are distinguished from one another not just because one of them (the Buddhist kind) is restricted to a particular subject matter – one's own experience – but because they have different purposes and they form part of two vastly different enterprises. Scientific observation has the aim of gathering information for use in conjunction with the two intellectual modes. Buddhist 'Bare Attention' – as also the Zen kind of attention to the here and now – is designed to bring the practitioner to enlightenment. It is generally agreed that the experience called 'enlightenment' cannot be adequately put into words. But, as we have seen, people do try. To use minimal and guarded language, it seems to be an expanded and highly satisfying state of consciousness.

In accounts of this experience from Buddhist and other sources (not all of them religious, as we shall see) the notion of transcendence and of some kind of state beyond the limits of space-time as ordinarily known to us occurs again and again. Keith Ward, speaking of the Buddhist notion, says:

Perhaps, then, the best way to think of it is not as self-negation nor yet as self-fulfilment as this is ordinarily understood, but as self-transcendence – finding one's truest reality in being fully attentive to the unconditioned, which brings bliss and knowledge.

Later he describes the experience thus: 'It is a state of having passed beyond the impermanent to an enduring centre.'[25] That is, it is a state which transcends time. (Note, however, that while Ward accepts the occurrence of such experiences he questions their use to justify claims about ultimate reality.)

We are brought to the conclusion that, within the Buddhist traditions we have been considering, escape from the line mode and the core construct mode mainly entails movement into two other modes: a special detached or objective kind of point mode, and a version of the transcendent mode. Characteristically the former is seen as leading on to the latter, which is the ultimate aim.

Further parallels between scientific method and Buddhist practice thus become evident. Both systems depend on detached, uninvolved, direct observation, used in conjunction with some transcendent function.

The analogy must not be pressed too far, for there are glaring differences. For example, science does not take the one mode as a path towards the other: it does not use objective data-gathering as a means of achieving *mathematical* insight. Yet the resemblances are interesting all the same. And they lead on, very obviously, to the question of the role of the construct mode. Does it have a part to play in the spiritual enterprise that is in any way comparable to that of the intellectual construct mode in science?

Before trying to answer this question it will be as well to broaden the discussion by an explicit consideration of the notion of meditation and what it is supposed to achieve, not only within Buddhism but elsewhere. Then the possible contributions of the construct mode may become more clear.[26]

Meditation has many different varieties, but all of them seem to have as their proximate aim the training and guiding of attention. When one meditates one keeps one's mind as steadily as possible on a single process – such as one's own breathing – or an object, or an image. This is not easy to do for more than a few minutes, as anyone who tries it soon discovers. Even if one reduces interruption from outside to a minimum by choosing a quiet place, by shutting one's eyes (if one is not meditating on a physical object) and by holding one's body still, one quickly becomes aware that thoughts and feelings do not readily stay still. There is, indeed, no better way to realize – if one did not know it already – the restless, shifting nature of one's own mind.

Thoughts and feelings wander. They wander in ways guided by what psychologists call 'associative processes'; and in the functioning of these processes the pull of the line mode and of the core construct mode is strong. So we keep coming back to personal concerns.

This is what the meditator is trying to learn to regulate. Most advice on how to meditate recognizes, however, that the mind is not at all easy to discipline. So the usual recommendation is to accept that intrusions will occur and, when they do, to note them then return gently to the meditation theme.

The difficulty of holding attention steady is recognized in a famous passage in the Bhagavad Gita where Arjuna says to Krishna: 'The mind is very restless, O Krishna, impetuous, powerful and firm. I think it is as hard to control as the wind.'[27]

With this, Krishna agrees; but he insists that control comes through 'practice and uncolouredness'. 'Uncolouredness' has to mean something like absence of emotion. The trouble is that achieving this is precisely part of the problem, for it is often emotion that pulls the mind away. However, with regular practice, most people seem to get better at keeping the attention steady. But what is the gain? Why do it?

It is at once evident that meditation is a technique for fostering the ability to move into a desired mode at will. Often this is the point mode, as in many Buddhist and Hindu practices. But it need not be. We saw earlier that St John of the Cross recommends meditation on 'forms, figures and images, imagined and fashioned by [the] senses', which is clearly a construct-mode activity. This, however, is for beginners, he says. Later comes the more difficult attempt at that movement into the value-sensing transcendent mode which John calls 'contemplation'.

We have already discussed at some length the Buddhist aim at reaching the transcendent mode; but in those varieties of Buddhism so far considered the use of images is not encouraged. However, within Buddhism, too, one finds disciplines where entry into the value-sensing construct mode is valued and cultivated. For example, the Yogācāra school favoured the idea of using meditation as a way of acquiring complete control of one's imagery.[28]

And imagery has a central part to play in the kindred Vajrayana or Tantric Buddhism that developed in Tibet. Within this tradition a distinction is drawn between the Path of Form and the Formless Path.[29] In the first of these there is a very great emphasis on the practice of elaborate visualization rites specified in considerable detail. But it must be noted that these are always explicitly a means to an end: the images are to be fully developed and then deliberately dissolved again.

The links with St John of the Cross are evident. In his view, too, images are to be cultivated then left behind, like a flight of stairs that serves only to give access to 'the level and peaceful room at the top' (see p. 153). And it is striking that St John talks of his recommended combination of construct- and transcendent-mode activities as offering 'a way to reach divine union quickly', while the Tantric Buddhists speak of their techniques as opening up a Short Path to liberation – assuming, that is, that the techniques are applied with enough resolve and energy. (By 'short' they mean liberation in a lifetime rather than in many lifetimes.) Now it is not thought impossible to advance swiftly without the use of visualization, but this is regarded as a much more difficult achievement. That is, the Formless Path – which in effect bypasses the construct mode – may also be short if one is highly adept and determined. However, it is generally believed that most people will make the best progress if they do not reject the Path of Form.

Many other examples could be cited from different traditions to show that meditation (using the word in the broad sense) may be practised as a technique for shifting into the point mode, or the value-sensing construct mode, or the value-sensing transcendent mode and for staying there during controlled and predetermined periods of time.

There can be no doubt that, among these three modes, the construct mode is the most controversial. Some have seen it as a help, some as a barrier to progress – something to be got rid of. But whatever view is taken of this, meditation is a technique for fostering by exercise the ability to regulate the locus of concern. It is a means of cultivating repertoire control.

This is the essence of its significance. Yet two further notions about its value are often met with, both of them related to the idea that it 'stills' or 'empties' the mind. The first of these is ancient in origin and amounts to a claim that when the mind is stilled or emptied – when the habitual chattering and fluttering die away – there may be room for 'something' to arise there instead – something greatly to be valued that would otherwise be crowded out. Usually this 'something' is considered to be an intimation of a reality beyond what we are ordinarily able to apprehend. The idea is therefore closely akin to the belief that meditation can help us achieve some kind of breakthrough from the finite to the infinite.

The second claim is very much more mundane. It is that meditation, by calming the nervous system, has physiological effects of a beneficial kind.[30] Notice, however, that the two claims are not at all incompatible.

Buddhism has served as an example of serious and sustained endeavour to learn how to diminish the role of the line and core construct modes in favour of some combination of the point mode, the value-sensing construct mode and the value-sensing transcendent mode. I want to conclude and summarize by focusing on some themes that have emerged and that are of quite general significance.

First, what has been called the 'intellectual path' – or sometimes 'the path of gnosis' – in meditative and religious disciplines the world over seems often to take as its starting-point a calm, attentive observation of one's own experience – a study that aims at some kind of objectivity and, indeed, impersonality. So the focus of concern is *me*, *here*, *now*. But this is not self-concern in any ordinary sense. The idea is that *this* kind of attentiveness to self leads precisely to loss of self-interest. The underlying belief, extremely widespread but varyingly expressed, is that human beings have the capacity, ordinarily dulled by our striving and fussing, to *see* (in some sense) into the heart of things by looking inwards. And this capacity, if properly developed, entails seeing ourselves in a new way.

That this emphasis was particularly strong in one, later suppressed, branch of the early Christian Church is shown by the writings known as the Gnostic Gospels that were found in 1945 at Nag Hammadi in Upper Egypt. For example, the Book of Thomas the Contender tells us that 'whoever has not known himself has known nothing, but he who has known himself has at the same time already achieved knowledge about the depths of all things'.[31]

So, if we consider our own experience in the proper detached point-mode way, waiting and observing, the belief is that we will apprehend reality more truly. Then we will discover that we are not as we thought we were; and in so doing we will lose the old phenomenal self and gain instead a glimpse, at least, of our true self, often written as 'Self'. This Self is said to be 'at one' with the universe, no longer separated and opposed to all else, that is, but somehow participant, transcendent, timeless. And the discovery of this is extremely powerful. Once we have even a taste, we want more. Wood says: 'When the Self is seen even a little, its appeal or pull is tremendous.'[32] That this is so is generally agreed by those who claim any direct experience of the kind we are considering.

Now it can scarcely be doubted that the experience, qua experience, is real and widespread. It cannot be called commonplace, yet it may not be so very unusual even today. This in itself, however, clearly does not tell us whether the claim that by this experience one comes closer to a knowledge of reality is well founded. The feeling of achieving this could be illusory. On the other hand, some of the greatest thinkers that humanity has so far produced have believed that it is rather our ordinary perceptions that are illusory. The outstanding case, perhaps, is Plato who, in the famous allegory of the cave, invites us to consider a race of beings chained in darkness from birth onwards, with fires burning behind them projecting shadows on to the cave wall. But even the shadows are shadows of puppets. What these prisoners see are thus only distant distortions of reality. And the prisoners can apprehend truth only if they are released from bondage, turned around and brought stage by stage (for it is painful) into the light of day.[33]

Plato is as convinced as any Buddhist about the unsatisfactory nature of our ordinary dealings with the transient and the perishable. His remedy, too, is to learn to contemplate the timeless, the transcendent. But his way towards this is somewhat different: it is to study pure mathematics. Thus Plato's 'way of the intellect' is indeed the way we now would call intellectual – which is hardly surprising since we owe to him so much of our intellectual heritage.

He says, for instance: 'For while they [mathematicians] employ by way of images those figures and diagrams aforesaid . . . they are really endeavouring to behold those abstractions which a person can see only with the eye of thought.'[34]

Thus the way towards an apprehension of reality is by pure reason. Plato's *ultimate* concern, however, is not with truth but with what he calls 'the Good'. And he says of the Good that it is the power 'which supplies the objects of real knowledge with the truth that is in them and which renders to him who knows them the faculty of knowing them'. Thus 'the Good, far from being identical with real existence, actually transcends it in dignity and power'.[35]

Recall that 'real existence' is already set in sharp contrast to all that we directly perceive and deal with in the transience of space-time. The Good is thus *very* remote and high and powerful. Plato's metaphor for it is the sun. It is the source of the light that enables us to see the real things lying outside the cave of shadows – and it is very hard to look upon directly. Commenting on the Platonic notion Iris Murdoch says:

The sun is seen at the end of a long quest which involves a re-orientation (the prisoners have to turn around) and an ascent. It is real, it is out there, but very distant. It gives light and energy and enables us to know truth. In its light we see the things of the world in their true relationships. Looking at it itself is supremely difficult and is unlike looking at things in its light. It is a different kind of thing from what it illuminates.[36]

Plato is evidently talking about spiritual aspiration. And the final object of spiritual vision he calls the Good, which implies that it is the ultimate value and so bound to evoke reverential emotion.

He believes, indeed, that to know the Good is to love the Good: if you do not love it, this is because you do not know it. So it seems he is claiming that the intellectual transcendent mode, in the form of mathematics, can serve as a path – *the* path, even – to the value-sensing transcendent mode in its most highly developed form.

Whether this could work for most people seems doubtful. However, Plato does also acknowledge that, in practice, the contact with changeless truth through insight into 'pure living mind' is, for incarnate beings, at best limited and occasional.[37] And he is conscious of the importance of keeping intellectual activity alive and vivid. What Whitehead would call 'inert ideas' are useless. Hence Plato is wary of the written word and believes it is good to look at the stars and *talk* philosophy. He is also wary of art, thinking of imagery as likely to make us feel content with appearance and as offering a false impression of transcendence.

So Plato probably has to be counted among those who see the construct mode as a barrier rather than an aid. Yet, as Iris Murdoch points out, he himself makes the most striking and effective use of images, as in the cave allegory itself.[38] Murdoch believes, as I do, that the value-sensing construct mode (though, of course, she does not use the term) can be a help rather than a hindrance. She defends art as giving us 'intermediate images' and argues, correctly I think, that most of us cannot do without the 'high substitute for the spiritual and the speculative life' that it provides. But she also recognizes that images can lead to a full stop if they are taken as being 'for real'.

What, then, in the end are we to think of as illusory? Is it the everyday world? Or is it the experience of transcending this world? Is it both – or neither? Whatever conclusion we reach, it would be arrogant to reject out of hand the claims of those who, like Plato, believe it possible to see truth and goodness more clearly than most of us do most of the time. At the very least some sort of humility in the face of such claims seems called for.

14

Value-sensing Experiences: Some Contemporary Evidence

It is time to come back to a consideration of modern Western societies. And the question that must now be asked is whether the sort of transcendent experience we have been discussing is known, in more than very exceptional cases, to their citizens. However, good evidence about this is not abundant. In such cultures people tend not to speak much of these things and the effort made to find out about their nature or prevalence has been slight in comparison with our attempts to gain knowledge of other kinds.

Still, there have been some serious pioneering investigations, particularly since the 1960s. These have been quite diverse in respect of the groups studied and the ways in which data have been collected. In some instances random sampling techniques were used. In others reports were obtained through mass-media requests for information. Sometimes special groups, such as nurses or graduate students, were questioned. Again, the questions that were asked varied considerably.

It must be said that the devising of suitable questions is not easy. An ideal question would be readily intelligible to most people yet not so formulated as to impose – or even suggest – particular ways of interpreting whatever experiences may have occurred. Both of these criteria are hard to satisfy, and researchers must have struggled with the task.

Here are some questions that have been tried:

'Would you say that you ever had a "religious or mystical experience", that is, a moment of sudden religious insight or awakening?'

'Have you been aware of or influenced by a presence or power, whether you call it God or not, which is different from your everyday self?'

'Have you ever felt as though you were very close to a spiritual force that seemed to lift you out of yourself?'

The last two of these are evidently quite similar, and they are quite specific as to the kind of experience being asked about. They try in different ways to avoid building in a firm assumption that the experience is an encounter with God.

These three questions have been used in a series of surveys in Britain and the United States. An excellent account of the work is given by David Hay.[1] There is a fourth question, somewhat different in kind, that I want also to cite. It is less specific; it suggests, I think, no religious assumptions, but it is probably harder to understand. It is:

'Do you know a sensation of transcendent ecstasy?'

This was devised and used by Marghanita Laski. If people asked what she meant by it, she answered: 'Take it to mean whatever you think it means.'[2]

Her subjects were the first sixty-three people that she met in circumstances which let her conduct her inquiry 'without intolerable embarrassment'. This meant in effect that they were mainly her friends and acquaintances, and so perhaps they were more likely than most to be able to cope with the interpretive problem. Certainly the accounts they gave contained enough consistency to show that their interpretations were not widely divergent.

However, if we do not restrict ourselves to a single study but try to take stock of the range of available research, then we find – unsurprisingly, given the differences in method – that the experiences quoted are quite varied in kind. There are, for instance, premonitions, appearances by dead people, out-of-body experiences, and so on. But also again and again one comes upon descriptions that satisfy the criteria for the value-sensing transcendent mode: the locus of concern does not lie in the events of space-time, and the predominant mental component is powerful emotion. This emotion is felt to be a response to something apprehended as being of intense reality and value. The value, however, is not specific to happenings in the individual's life, and so by contrast with it the ordinary self, its past and its future, cease to matter. The self may, indeed, for a moment cease to be sensed at all in its usual form, with the effect that hard boundaries of individuality and difference dissolve away.

A good example of a classic unitive experience is quoted by Hay:

I was walking across a field, turning my head to admire the Western sky and looking at a line of pine trees appearing as black velvet against a pink backdrop, turning to duck-egg blue/green overhead, as the sun set. Then it happened. It was as if a switch marked 'ego' was suddenly switched off. Consciousness expanded to include and *be* the previously observed. 'I' was the sunset and there was no 'I' experiencing 'it'. No more observer and observed. At the same time – eternity was 'born' – there was no past, no future, just an eternal now . . . Then I returned completely to normal consciousness, finding myself still walking across the field, in time, with a memory.[3]

Geoffrey Ahern, having carried out computer analyses of a sample of data collected by the Alister Hardy Research Centre, reports finding clusterings which suggest a major distinction between experiences that are 'spiritual/religious' (his term) and those that are not.[4] The reports in the former category tend to contain references to intensified perception (of colours, and sometimes of smells); to feelings of timelessness; to loss of the sense of separateness; to the gaining of a kind of insight that defies expression in words; to joy, bliss or peace. These, however, are not necessarily all present together.

Given the present state of knowledge, it is not possible to be sure how many people have these experiences today, but certainly the number is not vanishingly small. Most people who report such a happening say that it has come very rarely: once or twice in their lives maybe. And then it is usually brief, over in a moment. However, it is generally regarded afterwards as having been of great significance and value, though it is seldom spoken of. People keep it to themselves.[5]

The experience proves to be more common in some sections of the community than in others. Ahern reports a strong positive correlation with the level of education.[6] This may seem strange at first, since the intellect receives such emphasis in our present educational system; but it is not really hard to see how the finding makes sense.

Ahern himself argues that our inner cities have become particularly 'impermeable' to spiritual experience. It fits well with this that reports in his own study and in Laski's often mention as external triggers specific encounters with nature or art.[7] Also there are repeated suggestions that being alone is helpful or even necessary. The lives of the least privileged of our people are clearly in all these ways seriously short of opportunity.

Laski speaks of inhibiting circumstances, which she calls 'anti-triggers'. Commenting on their nature, she says:

Generally speaking they consist in anything that is inalienably associated with ordinary social life, and a useful general impression of them is gained by putting together what people in the groups felt they had escaped during their ecstatic experiences – ordinary everyday things, cares, commerce, conventionalities, ridiculous desires and purposes, worldly enjoyments, crowds.[8]

These ordinary everyday concerns are, of course, the province of the core modes in at least some of their manifestations.

The role of the intellectual modes as revealed by Laski's study is more complex. A few members of her questionnaire group cite an intellectual trigger – for instance, 'finding ten chromosomes when I knew they ought to be there'. This same person describes the experience of 'transcendent ecstasy' as 'a state of not being oneself'. Another gives 'solving mathematical problems' as one of a range of triggers and says of the ecstatic experience: 'Words like "liberation" – multiplication of possibilities – intense perception would go with it, hardly a part of it – confidence – a feeling of nearness to splendour.' Perhaps these people are, indeed, following the Platonic route and approaching the value-sensing transcendent mode by way of the intellect.

Laski comments that the pleasures of reason are still available to only a few people and that, even if reason can be itself a trigger, the exercise of analytical thought can also be 'irrelevant if not inimical' to the enjoyment of other triggers. She adds:

. . . in the very moment of ecstasy, the faculty of reasoning is suspended. What is lost in ecstasy comes to be judged as bad – and it is not possible to experience ecstasy and retain the power of reasoning, even though it may have been the exercise of reason that induced the ecstasy.

She claims, in other words, that even if intellectual transcendence can lead on to value-sensing transcendence the two cannot be experienced together.

We are now, I think, in a position to say that the asymmetry between the intellectual and the value-sensing modes in modern society is not inevitable. More than this: it may not be as great as it has been made to look by our conspiracy of concealment.

It is at any rate clear that experiences in the value-sensing transcendent mode are not extinct among us. They surge up still in spite of the power of other modes which have threatened to exclude them. However, there is one striking fact that I have not yet mentioned. In the great majority of cases people who report these experiences have not tried to cultivate them and do not claim to know how to do so. The experiences come occasionally, unexpectedly, like marvellous accidents; but they do not constitute a resource that can be used for living. They cannot be counted as part of the modal repertoire.

The question now is whether this might change. Is it possible within a society such as ours to learn how to foster the movement of the emotions beyond the line mode and the core construct mode so as to develop and cultivate these neglected capacities? Ahern would say that we do not just neglect them, we actively block and resist them.[9] And I think that, indeed, the current culture is in various ways not merely indifferent but hostile.

However, there may be signs that something new is beginning to happen. As I write, Ahern's study is very recent, while Laski's dates from thirty years ago. During this interval there seems to have arisen, in certain sections of society at least, a quickening sense of the need for change, a kind of hunger for it. This has led to proclamations that we are entering upon a new period of human life, sometimes called the Age of Aquarius, sometimes just the New Age.

New Age literature is now abundant. New Age events in the form of lectures, seminars and short courses are continually on offer. This morning's mail brought me a clutch of leaflets advertising such happenings. To take just a few examples, these proposed to teach me how to explore my inner potential, work with the wisdom of the body, restore disturbed equilibrium, release untapped sources of energy, learn ancient methods of shamanism, become a healer, control my dreams through secret techniques, use crystals to overcome illness, develop intuitive and creative thought, use colour and sound to expand awareness, discover how events in my past lives are affecting me now, meet my spirit guides, find my soul mate, dowse for health and learn the art of loving.

I do not mean to mock the urges which are seeking satisfaction in these ways. Nor do I by any means want to imply that all the lectures, courses and workshops offered are without value. Indeed, I am sure this is not so. But the range of options is bewildering, the claims are very great, and the dangers for the gullible are obvious. Also the goals themselves are quite heterogeneous. Many are really line-mode goals, having little to do with the cultivation of the value-sensing modes as I have here defined them.

I think it likely that this cultivation, like that of the intellectual modes, calls for steady work, sustained over many years; that it is hard to achieve in any circumstances; and that it is even harder to achieve without the full support of the resources of one's culture.

There exists an account of one serious and sustained attempt to cultivate the value-sensing modes that deserves attention here for what it shows to be possible. The attempt was begun by Marion Milner in 1926, when she was twenty-six years old. At that time there was, I think, no talk of a New Age to give encouragement;

and the general intellectual climate was probably more hostile to such an enterprise than it is today.

Marion Milner had what she describes as an ordinary Church of England upbringing followed by a university education. These left her, in her twenties, with an agnostic opinion about religion, a first-class honours degree in psychology and a general vague optimism about life, punctuated by outbursts of misery. Most of us would simply have continued in this state – 'making the best of it', as we say, which really means just enduring. But Milner took a different course which was to lead to developments initially quite unforeseen.

A critical turning-point seems to have come with the growth of her awareness of 'certain mental discomforts' which she had previously 'not known, only suffered'. She attributes this new acknowing to her habit of writing things down – scribbling random notes about whatever was on her mind. Looking over these later, she was amazed at the vehemence of the feelings they expressed; and thus she came to realize the extent of her dependence on what other people thought of her and the unease it brought her: the dread of giving offence or not doing the right thing, the inability to forget herself. Having once faced this unpleasant truth, she made up her mind that something had to be done. She took a decision that seems simple enough but proved to be momentous: she would try to find out what made her happy.

One might be tempted to make two dismissive comments: first that the answer is obvious, for surely one knows when one is happy; and then that the enterprise is selfish. But neither of these would be justified. Milner was certainly not just wondering how to have 'a good time', as we say. She wanted a guide for living – a way of knowing what matters; and she asked herself whether we may not perhaps be equipped with some intuitive sense of this that has been overlaid by social convention and dulled by neglect. If so, might we not then revive it by getting into the habit of attending to it? And wouldn't happiness be a sign that we were succeeding?

She did not begin with any conviction that this would work. But she thought it was at least worth a try. So she began to keep a diary in which she noted what she wanted, whether she got it and whether it made her happy – as well as 'anything else that seemed important so that if it should turn out that happiness did not matter, I should have a chance of finding out what was more important'.

Her first attempt to make an honest record for one ordinary working day filled her with such horror that she almost gave up. She discovered in herself so much concern about whether, for instance, her hair looked nice that she was dismayed and began to fear that the introspective diary-keeping would only make things worse. Yet something stopped her from trying to defeat her egotism by ignoring it. She could not give up the notion of a 'reality of feeling rather than of knowing' that had to be explored if one was ever to have 'a life of one's own'. This evokes in my mind strong echoes of Buddhist accounts of the search for 'insight', for by 'knowing' Milner clearly means intellectual knowing as understood in the Western tradition; and her 'reality of feeling' seems close to affective knowing or intuition. However, she gives no indication of recognizing this; and she expressly did not want to rely on traditional knowledge. So at the age of twenty-six she set out on her search alone.[10]

Milner herself thinks that the methods she developed may be of more general interest than the discoveries she made, since she wants to stress that what works for her may not suit everyone. However, in the present context both methods and outcomes are relevant. I shall try to summarize what emerges about them from the detail she provides.

The method has two parts. First there is the careful observation of one's own experience, including what Milner calls 'the small movements of the mind'. But then also there is the development of techniques for changing that experience. Both turn out to call for considerable honesty, doggedness and ingenuity.

The mind's 'small movements' were to prove particularly significant, for the recognition of them led Milner to see that it was possible to control the way she perceived things. It was after various deliberate experiments with these movements or 'internal gestures' that she had the sudden experience while watching gulls in flight that has already been quoted (see p. 192). In brief, what happened was that, quite unexpectedly, she saw the birds in a new way that changed idle boredom into 'deep-breathing peace and delight'.

It was then necessary for her to examine more closely the nature of this new way of perceiving, contrasting it with the old. After various attempts to arrive at the essence of the difference Milner expressed the distinction by using the metaphor of looking with a narrow focus and looking with a wide focus, the former being the way in which we perceive things most of the time. She puts it like this. Narrow-focus attention 'selects what serves its immediate interests and ignores the rest'. It is like 'a "questing beast" keeping its nose close down to the trail' – the trail, that is, of one's personal purposes. Wide attention, on the other hand, becomes possible only when these 'questing purposes' cease to be in control. 'To attend to something and yet want nothing from it, these seemed to be the essentials of the second way of perceiving.' (She emphasizes that this is not at all like vaguely preoccupied day-dreaming.)

It is at once evident what this means in terms of modes. Wide attention is a point-mode activity that is freed from any intrusion of purposes from other modes; but, more than this, it is freed even from the kind of purpose that is a component of most ordinary activity in the point mode itself.

Usually, as Milner says, if we want nothing from any object or situation we tend to ignore it. What she was beginning to discover was how to hold her attention wide open, to step clear of 'the distortions of . . . personal interests' and contemplate 'the calm impersonality' of things. She found that this brought delight of a kind she had not known before. And she began to practise what she calls the 'ritual gesture' of saying 'I want nothing', so that the experience could come to her.

This, however, did not always work. Sometimes the very desire for the experience seemed to prevent it, which is not so surprising since the desire is itself a personal purpose. And we have already recognized the paradoxical nature of desiring to escape desire. Sometimes, on the other hand, it seemed as if too much expectancy was the problem, so that 'vivid pictures of what might happen shut me off from perceiving what actually did happen'. Worst of all, however, was any fear of failure. Whenever the fear of making a mistake entered her mind, then her attention 'focused . . . to a pinpoint' and all widening became impossible.

By this time Milner felt she was genuinely making progress. The obstacles to further progress had therefore to be investigated. In particular, why was fear of failure so potent? This question led her to the recognition that much of her thinking was what she calls 'blind' – that is, given to uncritical beliefs, bad at weighing opinions rationally, bad at considering the points of view of other people and much entangled with emotion.

So now, as she began to understand this better, she again tried the act of detachment from personal purpose. Instead of asking how she could make people do as she wanted, she managed sometimes to ask: 'Why are they behaving like this?' More generally she tried to ask: 'What are the facts?' rather than: 'What shall I do?' And she speaks of the sense of relief and freedom which this shift brought to her. Blind thinking, she decided, cared little about the facts at all: 'Its ideas were guided by feelings, it believed on the whole just what it wanted to believe, though it always liked to pretend that it had taken the facts into account and acted reasonably.' She found, too, that its judgements were hardly ever moderate. It liked 'either . . . or' statements and tended to bolt to extremes. This last tendency, once she had become aware of it, helped her to understand the fear of failure. For if she had to be either marvellous or worthless, then a single mistake might reveal that the latter judgement was the right one. The awful truth about herself might be revealed.

Milner concluded that blind thinking went on like a kind of continual chattering at the back of her mind, destroying her peace. And, having noted this, she saw more and more clearly how certain kinds of emotion damage thought, pulling it down, reducing its potency.

By now, then, she was well aware of the trouble that can stem from the line mode and the core construct mode working together, especially when they work in ways dominated by the distorting demands of the self-image, as, in practice, they so often do. She concluded in the end that many of her problems stemmed from a very deep fear of loss of personal identity, and that this was what had caused her to be driven so hard by certain kinds of purpose: 'I had the desire always to be getting things done to prove to myself that I existed as a person at all.'

At about this stage in her inquiry she made a new discovery. She became convinced that her most elusive thoughts and needs – the ones she could not adequately put into words – were actually seeking expression for themselves. But, being wordless, they could find this expression only through images. She began, therefore, to try to use images as at first she had used language – that is, as a way of giving form to the vague contents of her mind. Thus images – visual mainly – became for her a new way to self-knowledge (and through the rest of her life, as her later books abundantly show, the way seems to have grown in significance).

From now on, Milner found that her moods came more readily under her control. But still not always. Sometimes tension and distraction were too strong. So she began to practise relaxation, and came through this to a new insight: she understood that the valuable kind of 'letting go' was not to be equated with complete mental passivity, for total passivity allowed her to fall back into 'blind thought' with all its hazards and inadequacies. In other words, some act of deliberate guidance, some steering of the attention, was needed if the tendency to slip into the line and core construct modes was to be avoided. Their pull remained strong. The mind was always apt to bolt along its old trails. She had therefore to cultivate 'that activity which was required to produce inactivity', namely a certain kind of mindfulness. The resemblance to Buddhist teaching is again evident, though Milner arrived at the notion in her own way.

At the end of *A Life of One's Own*, the book in which she describes these first stages of her explorations, she decided she had certainly confirmed her initial hunch that we *do* have an intuitive sense of how to live. But if we are to foster this and use it we must discover what interferes with it and how it can best be helped to find expression.

Milner has written other books in the course of her long life. One of these is specially relevant here as a sequel to the first but I shall mention it much more briefly.[11]

An Experiment in Leisure, published in 1937, deals with an uneasy worry left over from the earlier work: how do we really know when we can trust our feelings? To my mind this is a critically important question. One of the things which troubles me about some New Age claims − I mean specific claims about techniques like those illustrated earlier − is that the need to ask it seems seldom to be given prominence.

Because of her unease on this score Milner began to study some of her memories and to try to deal honestly with the darker side of her own nature. And much of the book is about the power of images. When are the pictures of the mind a strength, when are they a danger?

Milner concludes that the imagination can work in two ways: it can be the servant of fact − sometimes hard fact, so that through it we come to know ourselves better; or it can help us to evade fact by giving us substitute satisfactions when we are thwarted and frustrated. The critical question is: what makes for the shift between the one function and the other?

According to Milner the answer is to be found, once again, in the role of personal desires. If the imagination serves these − and particularly if it is used to glorify the self-image − then it is a danger. Otherwise it can be a great source of wisdom. For it can dramatically express truths about ourselves that we need to understand.

Her earlier general conclusion is thus confirmed: the repeated giving up of any personal purpose is a necessary condition for the growth and good order of the emotional life. And images can help this process, but only if they are properly used.

Milner is concerned in this book mainly with those pictures and metaphors that the mind throws up if one waits quietly and watches for them. But there is a similar distinction to be drawn in regard to public imagery, as she also clearly sees.[12]

The distinction depends, I believe, on the fact that images, like language, can be used in the service of different modes. In the functioning of the core construct mode they can readily further not just individual self-glorification but group purposes and power: the pure young hero going into battle, the utterly vicious enemy. They can then become, to quote Milner again, 'the instrument of the crudest infantile desire to be king of the castle and to prove that others are dirty rascals'.[13] So in the First World War British propaganda built up images of the incredibly cruel and barbarous Hun. And later the Nazis used pictures in which Jews were shown as ugly, loathsome, subhuman creatures, associated with dirt and vermin – while by contrast they promoted images of upright, beautiful Aryan boys and girls walking through sunlit meadows.

It is obvious that words can be handled to quite similar effect in some manifestations of the core construct mode. For instance, there are the simple little words 'red' and 'black' – and 'white' – that have filled so many hearts and minds with revulsion and dread. Or we might consider '*Deutschland über alles*' on the one hand and, on the other, that 'Land of hope and glory' which is to have bounds spreading 'wider still and wider' and is to become, if prayers are answered, 'mightier yet'. In these last two cases, and in very many more, music gives the words a great surge of extra emotional power.

There is nothing wrong at all with having – and expressing – a warm affectionate feeling for one's own people and one's homeland. We would all be deprived without this. The trouble arises when the positive emotions become linked to purposes which set the welfare of one human group against the welfare of others. The spreading of some territorial boundaries means, unfortunately, the shrinking of neighbouring ones. This is an inescapable fact about the world that we inhabit. To put it more generally, purposes, if they are of certain kinds, are bound to clash.

There is incompatibility that cannot be avoided; and so we are back where the argument began.

However, we are now in a position to compare this state of affairs with a different one; for no such conflicts arise when we succeed in moving into the intellectual or the value-sensing modes. If I increase my intellectual powers, no one else need suffer any corresponding loss. If I become better attuned to some 'reality of feeling', if I grow more emotionally responsive to values that transcend the personal, then no one else need pay a price for my gain. These are facts of the greatest importance for human lives and for the possible development of human societies.

15

Other and Better Desires: Prospects for a Dual Enlightenment

'Perhaps they will make me a king,' said Kim, serenely prepared for anything.
   'I will teach thee other and better desires upon the road,' said the lama in the voice of authority. 'Let us go to Benares.'
                                                          – Rudyard Kipling

Having tried to understand something of the scope for development of human minds, it is time now to turn firmly to questions of choice and action. These questions arise for us individually and collectively. We may answer them with varying degrees of reflection but we cannot avoid them.

   How, then, shall we use this dangerous, exciting capacity of ours to generate purposes and pursue them? What are the constraints – from within and without – that we must realistically reckon with? How hard are the enterprises that we might decide to set out on? How much help would we need, how much dedication? What do certain choices imply? What else might we have to give up? And, above all, what is worthwhile?

   We have already distinguished three ways in which a human mind can develop. There is expansion of the repertoire, that is, the addition of new modes to those already available. There is the learning of new kinds of competence within established modes. And there is the development of regulatory control of the repertoire, so that we become able to select the mode we want to be in at any given time, much as a driver selects a gear.

   Now it is evident that the range of possibilities which this affords is vast; and the first constraint is that no one can realize all of them. That is an unattainable goal. For one thing, the pursuit of excellence in a given direction generally takes a considerable amount of time. If you want to become a first-rate snooker player, there will not be much time left in which to acquire a high degree of specialized competence in mathematics. So you may have a great deal of mathematical potential which remains unrealized because you choose to spend your life in other ways. And even if, improbably, you managed to become both a great snooker player and a great mathematician, this would still be at the expense of other possibilities – infinitely many of them. Wittgenstein seems to have had much potential as an aeronautical engineer, but he chose to give preference to his potential for philosophy, sacrificing the other achievement.

All of existence is like this, as we have seen already. One thing comes into being and, by virtue of that very fact, others that were possible become 'what might have been'.

Thus even if it makes sense for us to think of ourselves as having some fixed potential, which is doubtful, we can never know what that is and we can certainly never fully realize it. So the idea is not useful. More than this – it can be very damaging if it leads us to decide that some people have a fixed potential that is low. If a child does not do well at school, what does that tell us about her potential for skiing or for spirituality, or even for intellectual development under other circumstances?[1]

Many people arrive early in life at the damaging, and often lasting, conviction that their potential, in some vague but quite general sense, is low. This is usually a consequence of accepting the way others see them. The value of thinking in terms not of limits but of possibilities is great, whether we are assessing ourselves or others; but it is surely at its greatest when we are assessing children. However, even this good policy, carried to extremes, has its dangers. Most policies, carried to extremes, do. We must always recognize that people differ: they have their own aptitudes. We are, as Whitehead puts it, 'naturally specialist'.[2] Also people relish different kinds of activity. One person likes being a television presenter. Another likes to sit in contemplative silence.

For each of us choice arises – choice as to what we will do and be. However, the range of options that is in practice open to us depends not only on ourselves but on our culture. Neither our own abilities nor our preferences develop in isolation from the other human beings among whom we live and grow. So while individual differences should always be respected, a number of questions arise for the conduct of society. Are there some kinds of competence that every member of the group should have? And, on the other hand, when it comes to specializing – as it always must – is there some kind of balance that should be aimed for?

What a culture encourages depends on many considerations and is not immutable, even though ideals, once established, may persist over long periods of time. We have seen how modes largely neglected for centuries may start to flourish again. Also, within modes, the most prized competences and the ways in which values find expression will differ from one place and time to another. The cave-dwellers painted on their walls, the Europeans of the Middle Ages built cathedrals.

If you live in a part of the world where the physical conditions are very difficult, then your social group is likely to concentrate on teaching you certain point-mode skills aimed at survival. But no society concerns itself with the point mode alone. All human beings develop a sense of personal time: memories for life events, anticipations of the future. These are laden with emotion. They affect what we think, how we feel, what we do. And then, beyond this, we also have emotion-laden beliefs about the general nature of things: about our personal identity and about our society, about the world around us, about the cosmos.

Every human society is concerned with such topics. And every human society expects of its adult members certain competences, certain kinds of skill and understanding, falling within the province of the three core modes: point, line and construct.

If you are to be a normally functioning grown-up human being anywhere, these modes must be in your repertoire, freely available for use. They are the ground modes; and they are universally mandatory. However, the precise kinds of competence that people are called on to develop within them may vary considerably from one social group to another. For example, many of us are now expected to learn to regulate our lives quite strictly in the line mode with the aid of our quartz watches. We must catch planes, keep appointments, turn up for work on time. In this respect we face, I think, unusually stringent and difficult demands; yet all societies expect people to be capable of some kind of forward planning and timing for participation in shared enterprises.

It is, however, the functioning of the core construct mode that seems to yield the greatest diversity among human groups. Yet while we are keenly aware of some of this diversity, often priding ourselves on our own ways of doing things, there are also certain of our constructed beliefs that seem to us so obvious, so absolutely 'given' as to be virtually beyond dispute. These constitute the realm of taken-for-granted know-how within which people conduct themselves sensibly, with 'common sense', as we say. And we tend to suppose that common sense is the possession of reasonable people everywhere.

I have spoken previously of 'human sense' as something that we do all hold in common.[3] But I now want to draw a distinction between human sense and common sense – a distinction which, in effect, defines the former notion more narrowly. Let us say, then, that human sense arises from early learning within the core point and line modes, learning that is similar in human societies everywhere; whereas common sense is amplified by later learning, particularly in the core construct mode, and is much more culturally variable.

Clifford Geertz gives an apt illustration of the sort of distinction I have in mind. He considers the 'common-sense opinion' that one ought to take shelter from rain if one can, and he divides this opinion into two constituent parts: the knowledge that rain will make us wet and the belief that this outcome is to be avoided. Then he points out that the knowledge part is universal: we all know that rain wets us. On the other hand, judgements about the desirability of taking shelter can be variable. It might be held that braving the elements, scorning the rain, is a good and proper kind of thing to do. Geertz might have added that, in parts of the world where rain is very scarce, one may rejoice in the feel of it when it comes.

Now knowing that rain makes us wet is part of human sense. However, judging what behaviour is appropriate in regard to that fact is part of common sense; and I agree with Geertz when he says this is socially constructed. At the same time it is experienced as so basic that it tends to assume for itself universality. As Geertz puts it: 'What simple wisdom has everywhere is the maddening air of simple wisdom with which it is uttered.'[4]

The fact is that all of us are expected to learn the common sense of our group in a very thorough kind of way, and mostly we do. If we do not, or if our learning falls into disuse, then we cannot meet the most fundamental expectations for sensible, not to say sensitive, everyday living. People who have spent long periods in institutions have great – and wholly understandable – difficulties with just this kind of coping when they emerge into society. They are often treated quite harshly in consequence. However, most societies have their special categories of tolerated incompetents: the unworldly priest, the absent-minded professor. We all know stories about such people. There was, for instance, a professor at St Andrews University of whom it is said that he travelled one day to Leuchars Junction, where he had to change trains, then phoned to ask his wife where he was supposed to be going. She told him, with her common sense operating, to look at his ticket; but he looked at the return half and travelled back to St Andrews.

So far, then, the argument is that the three core modes enter the repertoire readily and are mandatory for ordinary participation in all human societies. Some parts of their *content* – that is, the learning achieved within them – seem to be genuinely universal too. These constitute human sense. Some parts are so taken for granted in a given culture that they are assumed to be universal though they are not; and these are what we mean by common sense. There remains, however, a third component not yet considered: the core-mode learning of things that are culturally distinctive, recognized as such, prized as such and, if necessary, fiercely defended. I include in this last category a sense of our own history, as 'handed down' to us with its legendary elements, the old elaborated tales – sometimes songs – of triumphs and disasters. I include a sense of commitment to specific religions with their own festivals, ritual observances and sacred places. I include all norms for the conduct of life that we regard as giving us our national identity: this is how *we* do things, this is the way *we* are.

Now human groups are ordinarily concerned to preserve their own identities and perpetuate them. Thus, more or less deliberately, attempts are made to pass on these consciously held, distinctive beliefs and values to the young. Where the group is established and secure, these attempts generally have a good measure of success in spite of rebellions. But the case is different if an alien culture interferes, as has happened too often in history.

When one group manages to dominate another for any considerable period of time, the domination tends to have – perhaps *has* to have – as one of its main features the thwarting of the processes by which the subordinate culture is transmitted. Thus the dominated group is cut off, at least in some measure, from its own history and customs – and, critically, from its language – with resultant keenly felt loss and justified anger.

If we are to feel that we belong in a given social group then we must regard its distinctive core-mode beliefs and values as 'ours'. Loss of this comfortable feeling can come not only to members of subordinate groups but also to individuals living in a culture that is 'theirs' by other normal criteria. But more than loss of comfort is then at stake. If the alienation is deep and not well concealed it can be very dangerous for the person who experiences it. Persecution is a frequent consequence. For instance, disagreement with the dominant core construct system in Nazi Germany was perilous indeed. Alienation is usually risky in some degree.

The core construct mode is the arena for political and religious dispute. Words like 'heretic', 'dissident', 'subversive' and 'antisocial' are widely applied to those whose beliefs and values do not conform and who are seen as threatening. Abusive labels like 'scum' or depersonalizing ones like 'elements' may be added to make persecution seem more justified. 'Subversive elements' and 'anti-social scum' don't sound deserving of sympathy or gentle treatment.

Societies, however, can differ greatly in their tolerance for deviance and eccentricity. Also people who were at one time regarded as dangerous dissidents may acquire followers who venerate them. Christ was crucified for his deviance.

It appears, then, that the first kind of development, namely expansion of the modal repertoire, is common to everyone at least up to the core construct level. However, some kinds of development *within* these modes are quite variable, being affected both by individual predilections and by the influence of others. This influence is educational: the older members of the social group are teaching the young. But the teaching in question is often quite informal, not structured or planned. It arises spontaneously when a suitable moment presents itself (which is one reason why the total suppression of cultural traditions is hard to achieve). And frequently it comes in response to some initiative from the child – some interest shown, some question asked; though it is also true that many initiatives fail to evoke appropriate response and very many educational opportunities are not taken.

This kind of teaching has surely occurred in all human social groups that have ever existed. By its means, haphazard though it is, a great deal is learned in ways that seem to be efficient, enduring – and fairly pleasant as a rule, for teacher and for learner alike. The adult teaches when it seems appropriate, the child learns when spontaneous interest is aroused. This sounds – and is – good. But we come now to a question fundamental for education in a modern society: what about the development of the remaining modes, those that lie beyond the first three? Do they too become part of an individual's repertoire in the same easy, universal way? And, once this has happened, can learning *within* these modes be fostered effectively by means of the same kind of informal and opportunistic teaching?

I think it safe to say that there never has existed a human society all of whose members had all of the modes available to them, even if we set very low criteria for 'availability'. For example, many – perhaps most – of the adult citizens of a modern state have some ability to handle number in a way that depends on the functioning of the intellectual transcendent mode. That is, they can think about and operate on numbers to a limited extent without having to construct some imagined context of actual 'things' to be enumerated. So we might want to say that, as soon as this is possible, the intellectual transcendent mode is in their repertoire – is available to them, even if they are far from using this mode as freely as they constantly use the core three. But there are certainly some who leave school virtually unable to use the intellectual transcendent mode at all.

Similar arguments apply to the value-sensing modes, even in societies where these are highly valued. We have to conclude that, after the third mode, something changes. The core modes, it seems, 'come naturally' to human beings, but beyond this point we are in new territory, facing new educational problems. I stress again that to say these modes come naturally is *not* to say that they develop in ways independent of experience or of the influence of other people. Nor is it to say that all skills within them are equally easy and universal.

The core modes come naturally in the specific sense that all children who are neither severely abnormal genetically nor severely deprived or harmed environmentally will, by the age of four or five, have these modes available for free use and for much further optional expansion. This will have been achieved through sustained interaction with the physical and the social world, but yet without the need for consciously planned effort, either on the part of the child or of the community to which she belongs. Children learn a great deal that they never set out to learn and that no one else ever placed before them as a deliberately conceived goal.[5]

However, it is not the case that all human development can come in the same easy, unplanned way. Some achievements have to be won: won through sustained, deliberate practice or through structured study. We may call learning spontaneous when no one (learner or teacher, individual or social group) has consciously decided that the learning is to be undertaken. As soon as there arises a conscious decision to aspire to some skill or state of mind, or to get someone else to do so, then learning becomes deliberate. This is so even if attempts are made to make it appear spontaneous.

It has become fashionable lately, in Britain at least, to find these truths unpalatable. The idea that spontaneous, informal learning is the ideal model for all learning has proved powerfully attractive, to such an extent that many who really know better are afraid they will seem unenlightened if they propose that children should be systematically *taught* anything at all. This is death to the intellect. And, if the same doctrine is applied to value-sensing development it is, I think, equally destructive there.

The distinction between spontaneous and deliberate learning applies both within modes and between modes. Some kinds of point-mode skill come readily to everyone, some only with hard, deliberate work. We all walk but we are not all ballet dancers. Likewise, as we have seen, the core modes are freely and widely available, but this is not equally true of the remainder. In particular, it seems to be the case that to hold thought and emotion apart, even in a quite limited way, does not come spontaneously to us at all. If we are to learn to do it effectively, we must proceed in ways differing from those by which most of our early learning takes place. We must *apply ourselves*. We must become able to guide and direct our own minds.

Thus the need for discipline appears. And, though it is self-discipline that is in question, this is not easy to acquire unaided. Few can do it alone.

The question is: what help is needed and how can it best be offered? This question, so simple in appearance, is *the* educational question. The answering of it is peculiarly delicate and difficult. For there is a narrow path between the pitfalls that lie on either side.

In such a book as this only a few quite general points can

appropriately be made. The first two are about obstacles. We have to acknowledge from the start that the *sustained*, deliberate cultivation called for by the advanced modes may be rendered very hard, if not impossible, unless certain prior conditions are met. As I mentioned earlier, Whitehead found in mathematics an escape from 'the goading urgency of contingent happenings'. But he lived in highly privileged circumstances. Often the goading urgency is too strong to permit escape. It captures and holds us, especially if we are young and immature. How are children to undertake disciplined study if they are cold and hungry, or hot and thirsty, or have not slept well, or are frightened and abused? It is nothing short of cruelty to blame or scorn these children if they become 'educational failures'. Thus a certain level of material and social security is necessary. This applies to the intellectual and to the value-sensing modes in equal measure, so far as I can see.

Then there is the matter – the ever-recurring matter – of the self-image. Development beyond the core modes requires, by definition, cultivation of the ability to hold concern for the self-image in check, so that it can at least be laid aside for certain lengths of time. If these lengths of time are to be more than momentary, then some level of personal (not material) security may first have to be achieved. For when you are deeply and persistently worried about the 'sort of person' you are, concern about this will soon intrude; and instead of wanting to solve a problem you will be wanting to prove that you are the sort of person who can solve the problem. This is generally at the root of the fear of failure. And fear of failure is one thing that schools should be able to help children to overcome, though so often they do the opposite.

It must immediately be added that a child's sense of personal security does not depend only on what happens in school. It depends in profound ways on experiences in the family, where the self-image has its deep-reaching roots, and also on experiences in the wider social group. Evidently if a child belongs to a social subgroup which feels itself disparaged or threatened, then that child has a harder job to do in developing the kind of secure base from which extensive forays into the value-sensing or intellectual modes can best be made.[6]

I am reminded of the fact that a baby will generally explore a new environment more freely in the presence of a trusted mother to whom return is always possible. The very presence of the mother makes it easier not to bother about her for a while. It seems that a secure personal identity, a good-enough notion about who one is, facilitates the start of later mental explorations in a similar way. Understandably it is less alarming to stop bothering about the self-image – to forget about it temporarily – if one knows that one can turn around and find it reassuringly still there. Then gradually the importance of such a base may fade away; though it certainly does not always do so even when circumstances seem favourable.

Let us assume now – recognizing how utopian an assumption it is – that we are talking about a society where all has been done that can be done to remove those fundamental obstacles to progress beyond the core modes. And let us turn next to the question of how this progress can be positively fostered. What sort of conception of education will be most helpful in the deliberate cultivation of the intellectual and value-sensing modes?

It has come to be widely believed that the most enlightened kind of education is child-centred. This belief has found encouragement in certain arguments coming from within developmental psychology, most notably from the work of Piaget.[7] And it is a notion that appeals strongly to people of good will who want to protect children from the threats of oppression inherent in the opposite extreme, namely a culture-centred system which insists on compliance with cultural demands.

Now there is certainly a sense in which education ought to be child-centred. The teacher should always stand with the child – stand *by* the child in a shared enterprise, and try to understand imaginatively how it all seems from the child's point of view. Without this, how is it possible to appreciate the help that the child needs? Yet at the same time we have to acknowledge that the child's point of view is necessarily limited. In the nature of things children cannot envisage the possibilities open to them, or understand how these are to be realized. They cannot foresee the long consequences of moves made now. No one can do this fully, of course, but it is an essential part of the teacher's role to have likely long-term outcomes in mind, in ways that children are unable to do themselves.

Another way to put this is to say that the teacher must be able to take account of different points of view: the child's present point of view most certainly; but also the point of view of that same child looking back later at what her education offered or denied her; and the point of view of the cultural group, which has particular opportunities to provide and legitimate interests in what its members will become. I would go beyond even this and stress the importance of stepping back further still to recognize yet another point of view: the legitimate interest of all humanity. For what one group achieves or fails to achieve, values or fails to value, matters to us all.

So what we need is a kind of 'decentring' on the part of educators. I use the word quite strictly as Piaget used it. To decentre in his sense is to avoid being bound to a single point of view. It is to become able to appreciate different perspectives and see how they relate to one another. Young children can do this early and well in certain situations like the hiding games we discussed in chapter 6. But they cannot do it for themselves when it comes to the course of their education. How could they? How could they know enough in advance about the resources of their own society or about the wider resources of the human cultural heritage?

Thus both of the centred extremes in education – the child-centred and the culture-centred – have serious disadvantages. If education is culture-centred, then conformity, decorum and the conveying of information are overvalued, and there is an under-estimation of the child's ingenuity, imagination and initiative. There is also an underestimation of the human capacity for defiance and rejection and of the dire effects of boredom. Worst of all, perhaps, there is a strong likelihood that the children who are being taught will not be respected as human beings entitled to a say in the enterprise.

On the other hand, the main risk at the child-centred extreme is that of overestimating children's powers of self-direction and the validity of their judgements, while underestimating not only their need for systematic, well-thought-out help but also their willingness to receive this help if it is not forced upon them in insensitive ways. Young human beings have a remarkable fitness for the role of novice. They can enjoy accepting new goals and challenges from other people and can experience great satisfaction from the achieving of conscious mastery.

Good as children are at generating purposes, they must not be left to generate all of them alone. Nor do we serve them well if we habitually leave them to work out ways of achieving purposes for themselves. Children enjoy solving problems but, as Bruner puts it, 'they are not often either predisposed to or skilled in problem-*finding*' (my italics).[8] They like to make discoveries and it is good to challenge them to do so, but they also enjoy and benefit from having things explained to them. They ask questions and they want clear, honest answers, which, of course, if these are well-judged, need not close the inquiry but may provoke a further desire to know. Children are also capable of appreciating, at appropriate levels, the power and beauty of good *technique* – even perhaps of what Whitehead calls 'style', by which he means as 'the direct attainment of a foreseen end, simply and without waste'.[9]

A foreseen end, Whitehead says. But it is a fact of great educational relevance that ends differ from one another in the ease with which they can be seen from afar.

Some purposes are readily apprehended in their full scope even by quite young children. A child *can see what it is* to be a highly successful hunter and, in a more limited but still powerful way, what it is to be a dancer, a musician, or a painter. For example, a Chinese girl called Yani has recently achieved international fame for her amazingly beautiful and skilled paintings, produced between the ages of three and six. All of this began as Yani watched, and tried to copy, the activities of her artist father in his studio. Her father, suddenly seeing her interest, asked himself: does she really want to learn painting? The story goes that 'the idea seemed to lighten his mind. "Yani, you are really Papa's treasure," he said, kissing his daughter on the cheek because he was unable to hide his feeling of satisfaction.'[10]

From then on Yani seems to have received a very potent mixture of encouragement and sensitive help. Even when she ruined one of her father's works by smearing charcoal and paint all over it, she was not scolded; for she was accepting a purpose highly valued by her family and by her society.

The paintings Yani went on to produce remain her own. They are marvellously expressive of her enthusiasms and her humorous, playful interpretations. Yet they belong unmistakably to the traditions of her culture; and her achievements come from 'repeated and untiring practice' – more than 4,000 paintings in three years. So by the age of six, 'she has at least mastered some of the most important principles and skills of painting. Even the inscriptions and her signature are never carelessly thrown on paper, but are a product of careful consideration.'[11]

Yani received the purpose that started all this by watching her father. He did not deliberately plan to offer her this purpose, for she was only two years old. But without doubt he deliberately encouraged the subsequent processes by which this purpose became Yani's own.

This is one well-known way in which the direction of a life may be chosen: an admired and beloved person comes to serve as a model, willingly followed. But not all human activities are so readily displayed. What if the beloved person sits and thinks – or meditates – or prays? A child's understanding of what is then being done will not come so easily. And the activity, even when it begins to be understood, is not likely to seem so immediately appealing.

Also, if there are purposes that a culture deems highly important for many – or all – of its young people to adopt, then things cannot be left to chance. Such purposes must be proposed systematically. Encounter with them must not be merely fortuitous, dependent on a happy accident, like Kim's meeting with his lama (see the epigraph to this chapter).

Now no one can genuinely adopt a purpose without understanding it. It is true that you can perform actions without foreseeing the outcome to which they tend, and in this way you can uncomprehendingly further someone else's purpose. But you cannot make a purpose your own unless you can in some way foresee a goal. Hence, of course, comes the value of a visible model, an outcome displayed.

I take it as a basic tenet that education fails unless the purposes which it offers are genuinely accepted – that is, taken over by the learners as their own in the end. Otherwise discipline cannot become self-discipline. Consequently special educational effort must be devoted to making comprehensible those purposes that are most likely to seem obscure.

That people should not be forced into activities which seem pointless to them is, I am sure, the intuition underlying the claim that all education must be relevant to pupils' lives. And this basic intuition is sound. But education is about *changing* lives – about enlarging the scope of relevance. It is about increasing the modal repertoire, for one thing. It is about suggesting new directions in which lives may go. If education is so scaled down that it merely serves pre-existing purposes – purposes that would be there without it anyway – then this is just another kind of educational failure.

How, then, do we help children to understand goals that are quite far from their own present lives and not readily displayed? How do we give them any notion of what it would be like to achieve power as a mathematician, say, or as any kind of theoretical thinker? Or, again, how do we give them some sense of the experience that comes with developing spirituality as it aspires towards transcendence? The realistic answer has to be, I think, that these goals, in the fullness of their maturity, cannot be presented in any explicit form that young children can understand.

So what is to be done by a culture that wants to propose high levels of intellectual or value-sensing attainment to its young members? How can such a culture prevail on the young to make the kind of sustained, disciplined effort that is certainly going to be needed?

There are at least two things that can make a difference. First, if we are wise enough and sufficiently serious about the enterprise, our schools can offer intermediate goals in a well-planned sequence so that each achievement is also an opening which reveals new challenges not too far out of reach. The sequence, though, must not be rigid, still less rigidly enforced, and must never come to resemble in practice that dread image of hoops to be jumped through. The teacher is not a ring-master. Neither, however, is the teacher merely a consultant. The teacher is one who knows what lies ahead and also *how to get there*. The teacher has new desires to offer – and new techniques for learning how to satisfy them.

But this is not all. In spite of the difficulty of understanding really remote goals, there is something else that a culture can do to propose them to the young – something more pervasive and in a way more subtle.

Consider for a moment the traditional ways of the East when it comes to the offering of spiritual goals. Here the key is the figure of the holy man. Think of how this figure must appear to a child. The holy man does not work as others do. He is not a farmer, or a shoemaker or a merchant. He makes no obvious contribution to social life at all. Yet if he appears in a village with his begging-bowl people gain 'merit' by giving him some of their often scarce food. And if he decides to settle in a cave nearby they are pleased, even though the holy man will now need to be fed continually. A child who observes this perceives that the holy man is valued. She comes to understand that this person, for whatever strange reason, is to be treated in a special way: he is to be treated with reverence. Aside from the encounter with reverence itself, which is no small matter, perhaps the most important thing about this experience is to be found in the very fact that the reasons for the reverence are not immediately obvious. Thus the child's current values are challenged and the possibility is opened up of aspirations that lie beyond.

Jean-Paul Sartre, writing of his childhood, gives this account of his early encounter with books in his grandfather's study:

Even before I could read, I already revered these raised stones; upright or leaning, wedged together like bricks on the library shelves or nobly spaced like avenues of dolmens, I felt that our family prosperity depended on them ... I used to touch them in secret to honour my hands with their dust but I did not have much idea what to do with them and each day I was present at ceremonies whose meaning escaped me: my grandfather – so clumsy, normally, that my grandmother buttoned his gloves for him – handled these cultural objects with the dexterity of an officiating priest.[12]

Kim's first dealings with the lama are of a different order, and yet there is a deep underlying similarity. When he sets out on his journey as the old man's disciple, Kim has his own goals, stemming from confused memories of things his father told him. He treats these as prophesies and he wonders, idly enough, if his destiny is to be a king. But the lama promises to teach him *other and better* desires – not in the first place how to attain goals but how to have new kinds of goal, hitherto unimagined and so at first incomprehensible.[13]

A desire to exchange old goals for new is not the reason why Kim, the streetwise urchin, goes off with the old lama. He is play-acting at first in his role as the disciple. But the lama's invitation is the classic one. And since the new goals cannot initially be understood by the novice, the first step on the Way has to be an act of trust: trust that the guru will in the end have something to reveal that is worth the effort. This trust does not stem only from the person of the teacher as perceived by the learner, though this matters. The trust is encouraged and endorsed – or otherwise – by attitudes that are widespread in a society.

Society conveys a general message by showing respect to certain of its members who are adept at highly valued skills. When respect is shown for attainments not easily put on display, young people will at least be helped to understand the limitations of their own understanding. They will realize that they might aspire to something that is at present beyond them. And some will choose to try.

There are dangers in all this, and they are well enough known. If, for instance, representatives of a religion are accorded automatic respect, some – many, perhaps – will not merit it as individuals. They may even be fraudulent scoundrels. The same is true of those who represent officially the achievements of the intellect. Some do little credit to the values they are supposed to embody. However, my general point is unaffected, and it applies to the intellectual as much as to the spiritual: if children are to be encouraged to direct their efforts towards achievement in these modes, they need to be shown that high proficiency in them is valued. There may be different ways of demonstrating this; but if some good way is not found within any given social group, then in that group the modes in question will wane. Remember, these modes do not come to us spontaneously. They have to be cultivated by deliberate effort or we lose them.

This argument does not tend to the conclusion that skills in other modes should be valued less. There are many ways of contributing to the common good and I am not trying to put them in order of importance. I want simply to point out that some kinds of competence are much more visible than others; and that visibility is not a safe measure of value.

During the last century or so a number of social groups have tried to do something never attempted before in the course of human history: they have tried to introduce every one of their members to the intellectual modes. We tend now to take universal schooling somewhat for granted, but we should recognize the novelty and the difficulty of the enterprise.

Of course, the aim is not that everyone should become highly competent intellectually. That would take too much time and preclude the realization of other kinds of potential. This much is generally acknowledged even by those who value the intellect highly. However, there is wide agreement that universal schooling is meant to foster at least the fundamental intellectual skills: whatever else is to be achieved, children are to become reasonably literate and numerate, to know about some of the achievements of science – perhaps even a little about its methods – and to learn at least something about rational thought as applied to general themes. Where all of this is enforced by law it might perhaps appear that society gives the crucial message of respect quite clearly. Unfortunately, however, the message is often very far from clear. It is entirely possible for people to demand that children be taught literacy and numeracy and yet to have no understanding of the real nature of the intellectual modes – even to dislike or fear intellectuals.

The intellectual competences are commonly valued for purely line-mode reasons – that is to say, for their practical usefulness, which turns out to be great. Parents often want their children to acquire intellectual skills because thus they will 'get on'. Long ago in ancient Egypt, it was already acknowledged that 'the scribe is the boss'.[14] It still tends to be true. And recognition of this fact need carry with it no understanding of the special significance of literacy for the development of the mind.[15]

The goals of the intellectual modes, as of the value-sensing ones, are the transformation of the self in chosen directions. In each case there is a Way to be followed. In each case there is a need for years of discipline and dedication, with skilled help for the beginner from someone who knows the path and can propose intermediate goals that are near enough to be understood and to be seen as attainable.

I have argued that intellectual competence is not widely understood or valued for what it is; and that this is evident in attitudes to education. But the case is much worse when we turn to a consideration of the advanced value-sensing modes. Here the lack of understanding is pervasive and deprivation has become acute, at any rate for large sections of our society, and particularly in the 'impermeable' inner cities, to use Ahern's word again.

Many people feel the deprivation. Many more seem either not to feel it or else not at all to recognize the nature of what they feel themselves to lack.

However, there are certainly serious thinkers in growing number

264 *Range and Balance in Maturity*

who seem to have reached the conclusion that the Enlightenment of eighteenth-century Europe has failed us and that it is time for something new. Against this one might want to argue that it is rather we who have failed the Enlightenment – failed to live up to its highest principles and so failed to realize its promise. It may be true, as Stuart Hughes says, that the Enlightenment in its origins was less unbalanced – less heavily weighed towards the intellect – than it became after later distortions. Nevertheless, even as Stuart Hughes argues this he is aligning himself emphatically with the traditional conception of the Enlightenment as an age that held reason and science to be 'man's highest faculty'.[16] Thus our value-sensing capacities are being put quite firmly into second place.

What so often goes wrong, I think, is that advanced-mode thought is compared with core-mode emotion. The very possibility of emotional development that is genuinely on a par with – as high as, level with – the development of reason is only seldom entertained. So long as this possibility is neglected, then if reason by itself is sensed as inadequate where else can one go but back? Thus there arises a regressive tendency, a desire to reject reason and all that was best in the Enlightenment, a yearning for some return to the mythic, the magical, the marvellous in old senses of these terms. This is very dangerous; but it has the advantage that it is altogether easier than trying to move forward into something genuinely new.

Now we have clearly seen that the cultivation of the advanced value-sensing modes is not of itself new. It has ancient roots. What *would* be new would be a culture where both kinds of enlightenment were respected and cultivated together. Is there any prospect that a new age of this kind might be dawning?

I would like to think so; but how is one to tell? One may inquire about general preconditions for this kind of great change, but that is a vast and difficult question. Fernand Braudel, the distinguished French historian, stresses that we must never overlook the effect of what he calls '*la longue durée*'. In other words, we must recognize how greatly we are influenced – weighed down or boxed in – by the vast stretches of past time. *La longue durée* is a slow-moving affair, where one needs to consider a span of centuries to see any change at all. And Braudel argues that we are barely conscious of the way in which this kind of long-standing historical influence affects us now.

Yet sharp changes do occur, and Braudel recognizes this too. He recognizes the dramatic nature of the change that took place in the eighteenth century, and the fact that we are now living in a time which 'breaks the old cycles and the traditional customs of man'.[17]

So what will happen next? Among all the influences that will work together and interweave, some no doubt quite elude our awareness now. But one thing that makes a difference is the mental equipment available to people: the concepts they can use, the distinctions they are able to draw.[18] For example, consider the widespread failure in the seventeenth century to discriminate between mathematics and mechanics on the one hand, and magic on the other. We saw in chapter 10 how, because of this, techno-logical developments made possible by the advent of science were often taken to be supernatural wonders, stemming from the power of the devil, not from the power of human minds.

This particular confusion can only seem absurd to us as we look back across the intervening centuries. But we ought to ask ourselves if there may not be other confusions prevalent now that are just as damaging. And I think indeed that value-sensing transcendent experiences of the kind we have been defining and discussing are in great danger at present of being confused with other things to which they bear little resemblance.

There is a sort of double risk. On the one hand, they in their turn may be confused with magic, so that, as Paolo Rossi puts it, 'a return to the archaic phase of magical experience is accepted by many as a valid method of freeing ourselves from the sins of our civilization'.[19] Like Rossi, I regard this as a flight from our historical responsibilities.

On the other hand, experiences in the value-sensing modes run the risk of being confused with madness. Where value-sensing experiences approach mysticism, as the states we have been considering certainly do, many people hearing about them tend to feel acute suspicion which often expresses itself in ridicule and scorn. Those who admit – or aspire – to such experiences are in danger of being regarded as 'weirdos' or 'nutters'. The danger is particularly great, I think, if the experiences are not expressly linked to any orthodox system of theological belief. For such a system may seem to confer a kind of respectability.

Of course, there are people among us who are prone to various kinds of delusional state – and there are others who take advantage of these people for their own ends. Likewise there really were – and are – practitioners of magic.

When it was mathematics that had to be distinguished from magic, this was not easy, as we have seen. However, it was achieved. For our part we shall have to achieve a similar distinguishing of experiences in the value-sensing modes from magic on the one hand and madness on the other if we are ever to correct the imbalance between intellectual and emotional development that exists today.

My final word has to be that I do not know whether a new age, characterized by a double enlightenment, is anywhere near. But I am sure that the chances of its coming are enhanced if we add to our mental equipment a better understanding of how the core modes relate to the advanced modes; and also of how, within the latter category, the intellectual and the value-sensing varieties of experience relate to one another. We may then more easily see that, if the intellect has unbalanced us, there are corrective steps open to us which are not regressive and which do not entail a rejection of reason. At the same time, we may come to feel less embarrassed about and suspicious of transcendent emotion, seeing it as no more 'weird' than the capacity for mathematical thought. Neither of these is, or is ever likely to seem, banal or commonplace. Each has its element of mystery. Yet each is a normal, though generally ill-developed, power of the human mind.

Appendix: The Modes

The four main modes are defined by four loci of concern, as follows:

First, we may be concerned with something lying in that chunk of space-time we call 'here and now'.

Second, we may be concerned with events that have happened in the past or that might happen in the future: that is, with the 'there and then'.

In these two loci, concern is with specific things. However, as we move on, there is a shift in the direction of generality: the mind starts to be concerned with the way things are.

In the third locus, concern is with how things are 'anywhere, anytime' – or at least 'somewhere, sometime'. There is an attempt to deal, in some way or other, with the *nature* of happenings in space-time. Interest no longer centres on particular episodes except in so far as they illustrate some generality.

Finally, we come to the fourth locus. And perhaps now the term 'locus' is scarcely appropriate, for concern is no longer fixed on happenings in space-time at all. We may say, oddly enough, that concern is 'nowhere'. But that certainly does not mean the mind is concerned with nothing. The activities in this mode are very varied, but the most helpful example to give is probably mathematics. Mathematics deals with patterns of relationship. These patterns may have applications in space-time. Often – perhaps surprisingly often – they do. But not always. To take the most obvious example, there may be extension into n dimensions. But in any case, when we are engaged in pure mathematics our *concern* is with the patterns, not with their possible spatio-temporal instantiations. Thus the 'locus of concern' does not lie in space and time.

The names of the four main modes, as defined by the loci of concern, are:

point mode – locus 'here and now'
line mode – locus 'there and then'
construct mode – locus 'somewhere/sometime' (no specific place or
    time)
transcendent mode – locus 'nowhere' (that is, not in space-time)

There are also four mental 'components' to take into account, namely: perception, action, emotion and thought. These serve to specify subdivisions of the four main modes. So we have in effect a two-way table, with 'locus of concern' along one side and 'components' along the other. Four loci of concern and four components would yield sixteen cells, but things are more complicated than that because we have to allow for combinations of components. Four components – taken all together, any three, any two or singly – yield fifteen combinations. But even this does not adequately acknowledge all the subtleties that arise in practice; for the mind may sometimes function in such a manner that one component is dominant while another has a restricted, yet still important, part to play. Thus we cannot just record 'x and y' in any tabulation but must also be prepared to distinguish 'x and (y)' from 'y and (x)', where the bracket indicates a subsidiary role. For example, there is a kind of thinking, commonly called 'dispassion-ate', in which we aim to engage when we tackle certain kinds of problem. However, it turns out that emotion is not wholly excluded from this activity.

The table, then, has in principle four entries along one side and more than fifteen along the other. But many of the possibilities provided remain unused in practice. Some are logically impos-sible.

A restricted table showing the modes that receive most attention in the book given on p. 269 (see 'A Table of the Modes'). However, reference is also made at some points in the discussion

Locus of Concern			
Here Now	There Then	Somewhere Sometime	Out of Space–Time

Components		Here Now	There Then	Somewhere Sometime	Out of Space–Time
	Perception Action Thought Emotion	Point			
	Thought Emotion		Line	Core Construct	
	Thought (and Emotion)			Intellectual Construct	Intellectual Transcendent
	Emotion (and Thought)			Value–sensing Construct	Value–sensing Transcendent

A Table of the Modes

to certain variants of the point mode not shown in the table, such as that which is entailed in the attempt at detached or 'objective' observation. In practice many variants of the point mode probably occur.

It is evident, too, that there exist other variants of the line mode, for one can think unemotionally about specific events in the past or the future. For example, I can bring to mind, if I choose, the departure yesterday of the first morning flight from Edinburgh to London. This event has no significance for me, and the thought is accompanied by no emotion. What we call 'small talk' is generally in this unemotional kind of line mode. For example, one may ask casually: 'What did you do yesterday evening?' But conversation of this kind readily turns into gossip. The answer might be: 'Well, we had dinner at Valentino's and you'll never guess who we saw sitting at a corner table!' Then emotion of one kind or another is apt to come back in.

There is a second way of representing diagrammatically those modes with which the book is chiefly concerned (see 'A Map of

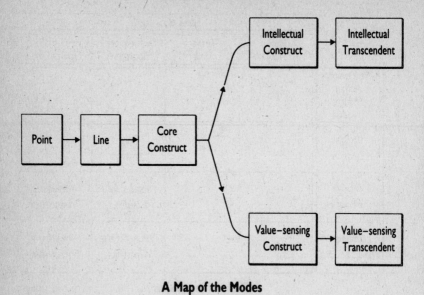

A Map of the Modes

the Modes', above). This has the merit of showing the direction of development from infancy onwards. Also it marks clearly the very important bifurcation that takes place beyond the core construct mode, at which stage the achieving of some measure of separation between thinking and emotion becomes possible.

However, if this map is used, it must be borne in mind that the arrows do not indicate the replacement of one mode by the next as development proceeds. Nor is the bifurcation to be thought of as a point of necessary choice – an unavoidable parting of the ways.

The three modes preceding the bifurcation are known collectively as the core modes. The others are called the advanced modes.

Notes

PART ONE: DEVELOPMENT IN CHILDHOOD

1   Modes of Mind: An Introduction

1. The exact quotation is:

   > Then close the valves of her attention
   > Like stone.

   It comes from the poem that begins: 'The soul selects her own society . . .'
2. A. Einstein, *Autobiographical Notes*, La Salle and Chicago, Illinois: Open Court, 1979. (First published 1949.) Einstein concluded that, moving with the beam of light at the velocity of light in a vacuum, he should observe it 'as a spatially oscillatory electromagnetic field at rest'. However, the puzzle was that there seemed to be no such thing. In this paradox lay the germ of the special theory of relativity.
3. I wrote this quotation in a notebook many years ago but, unwisely, did not record the source and now have failed to find it. However, I think I would have noted it accurately; and it is certainly the sort of thing that Einstein would say of himself.
4. The term 'intentionality' was brought into use again in the nineteenth century by Franz Brentano, who regarded intentionality as the distinguishing mark of mental phenomena: that which sets the mental apart from the physical. Brentano's principal work was called *Psychologie vom empirischen Standpunkt* (Psychology from the Empirical Point of View) and was published in Leipzig in 1874.
5. Reasons for thinking that this mode is not available until some time after birth will be considered in chapter 4.
6. This should not be taken to mean that independent functioning is

inherently superior to integrated functioning. What is acquired is a greater range of options.

7. R. Lazarus, 'The Self-Regulation of Emotion' in L. Levi (ed.), *Emotions: Their Parameters and Measurement*, New York: Raven Press, 1975. When Richard Lazarus speaks of 'coping' he means not only direct action on the outer world but also action aimed at the regulation of one's own mind. He discusses what he calls 'denial-related' coping devices used by air crews on combat flights – for instance, the creation of fictions of invulnerability. There will be more to say on this theme in later chapters.

8. A very interesting account of the development of emotions and of the levels of complexity that can arise is given in K. W. Fischer, P. R. Shaver and P. Carnochan, 'How Emotions Develop and How They Organize Development', *Cognition and Emotion*, 1990, 4, pp. 81–127.

9. Modern developments in telecommunication obviously introduce complications. It remains true that when concern shifts from present to past or future events, it is ordinarily the case that perception and action no longer play a direct rôle.

10. M. Donaldson, *Children's Minds*, London: Fontana, 1978.

2   Some Human Ways of Dealing with Hard Fact

1. Jean Piaget has made this claim in very many publications. J. Piaget and B. Inhelder, *The Psychology of the Child* (London: Routledge and Kegan Paul, 1969; first published in French 1966) provides an introduction to his work and an overview of his theoretical position; J. Piaget, *Logic and Psychology* (Manchester: The University Press, 1953) gives much more detail as to the postulated nature of the cognitive structures; and J. Piaget, *Biology and Knowledge* (Edinburgh: Edinburgh University Press, and Chicago: Chicago University Press, 1971; first published in French 1967) gives an account of the way in which the structures are believed to be reconstructed repeatedly as development proceeds. An appendix in M. Donaldson, *Children's Minds* (London: Fontana, 1978) attempts a brief summary of Piaget's theory of intellectual development.

2. E. Fromm, *To Have or to Be?*, London: Jonathan Cape, 1978.

3. Tversky and Kahneman discuss this in terms of the 'availability' of

examples of danger. See A. Tversky and D. Kahneman, 'Availability:
A Heuristic for Judging Frequency and Probability', *Cognitive Psychology*, 1973, 5, pp. 207–32.
4. For example, some people have certain specific fears or phobias that
they cannot deal with in this way. Phobias are usually fears of things
not in themselves dangerous, like open spaces, enclosed spaces, or
ordinary little spiders. Discussion of the nature and cause of phobias
would carry us far from the main course of the argument.
5. L. Weiskrantz, *Blindsight: A Case Study and Implications*, Oxford:
Oxford University Press, 1986.
6. One may ask whether the processes by which the status of some bit
of knowledge changes, so that it comes to be 'acknown', are
themselves conscious or not. I do not know the answer, but it seems
likely that often they are not in full awareness. Generally speaking,
we are more aware of the products of thought than of its processes.
   In an important paper Robin Campbell discusses the distinction
between those mental structures and processes which are accessible to
consciousness and those which are not. The former he calls *phenic*,
the latter *cryptic*. These are good words for the distinction Campbell
is making and for the purposes of his argument. They would be less
appropriate here because I want also to take into account knowledge
that is accessible but not always accessed. See R. N. Campbell,
'Language Acquisition and Cognition' in P. Fletcher and M. Garman
(eds.), *Language Acquisition*, Cambridge: Cambridge University Press,
1986 (2nd edition).
7. Freud used the words 'repression' and 'defence' in different ways at
different times as his theories developed and changed. Sometimes he
used them as synonyms, sometimes to make certain distinctions; but
for present purposes the distinctions are unimportant. A useful
survey of this topic is provided in P. Madison, *Freud's Concept
of Repression and Defence: Its Theoretical and Observational Language*, Minneapolis: University of Minnesota Press, 1961. See also
H. Sjöbäck, *The Psychoanalytic Theory of Defensive Processes*, New
York: Wiley, 1973. Hans Sjöbäck gives a very detailed account of the
writings of a number of Freudian theorists on that subject.
8. D. N. Stern, *The Interpersonal World of the Infant*, New York:
Basic Books, 1985. Nevertheless, there may still occur early precur-
sors of defensive manoeuvres at times when reality seems highly

unsatisfactory, for example, when communication with the mother is seriously disturbed. Lynne Murray (personal communication) makes the point that in these circumstances babies may withdraw into self-absorption.

9. S. Freud, *Analysis Terminable and Interminable*, 1937, Standard Edition, vol. 23.

10. Jerome Bruner has many very interesting things to say about the manner in which we construct 'more or less coherent autobiographies centered around a Self acting more or less purposefully in a social world'. He points out that families, and whole cultures, invent traditions in essentially similar ways, and speaks of 'the push for connectivity'. See J. S. Bruner, 'The Narrative Construction of "Reality"', closing address to the Fourth European Conference on Developmental Psychology, Stirling, Scotland, August 1990.

11. Having studied the topic experimentally, Harris concludes that young children make the reality-fantasy distinction yet still wonder if what they have knowingly imagined might be real. See P. L. Harris, E. Brown, C. Marriott, S. Whittall and S. Harmer, 'Monsters, Ghosts and Witches: Testing the Limits of the Fantasy-Reality Distinction in Young Children', *British Journal of Developmental Psychology*, 1991, 9, pp. 105–23.

12. I heard Bruno Bettelheim talking about this boy. I have not been able to find an account of the case in his published works.

13. S. Freud, *Introductory Lectures on Psycho-Analysis*, 1916, Standard Edition, vol. 16, p. 370. Freud claims that seduction, when it really occurs, is more often by another child than by an adult. He is at pains to exonerate fathers, saying: 'if in the case of girls . . . their father figures fairly regularly as the seducer, there can be no doubt . . . of the imaginary nature of the accusation'. Then a few sentences later he allows that sexual abuse of a child by its nearest male relatives does sometimes occur – but it will have happened in the later years of childhood, he claims, even if the child has 'transposed' it into earlier times. His general claim is that something deep in our instinctual nature, stemming perhaps from prehistoric truth', *demands* these events in childhood. He goes so far as to say: 'If they have occurred in reality, *so much to the good* [my italics]; but if they have been withheld by reality, they are put together from hints and supplemented by phantasy.' I have not read the original German,

but Freud seems here to offer extraordinary encouragement and justification to abusers.

14. J. M. Masson, *The Assault on Truth: Freud's Suppression of the Seduction Theory*, Harmondsworth, Middlesex: Penguin, 1985, p. 133. (First published, 1984.) Jeffrey Masson has been fiercely attacked for this book. I do not think I am in a position to judge whether he is wholly fair to Freud. He draws on unpublished documents to which I have no access.

3   The Point Mode and Its Origins in Infancy

1. This needs to be qualified by the recognition that the close scrutiny of unclear cases may sometimes be critical in the formation of theory. I am grateful to Robin Campbell for pointing this out to me.

2. R. L. Fantz, 'Pattern Vision in Young Infants', *Psychological Review*, 1958, 8, pp. 43–7.

3. M. H. Bornstein, 'Brain Mechanisms and Infant Visual Attention', *Journal of Experimental Child Psychology*, 1978, 26, pp. 174–92.

4. E. J. Gibson, 'Exploratory Behavior in the Development of Perceiving, Acting and the Acquiring of Knowledge', *Annual Review of Psychology*, 1988, 39, pp. 1–41.

5. An account of this work is given in J. S. Bruner and B. M. Bruner, 'On Voluntary Action and Its Hierarchical Structure', *International Journal of Psychology*, 1968, 3, pp. 239–55. In this connection we should note the claims made by Nicholas Bernstein about the very widespread occurrence in animal life of what he calls 'the modelling of future requirements' by the brain. (N. Bernstein, *The Co-ordination and Regulation of Movements*, Oxford: Pergamon, 1967.) In other words, we should not take it for granted that representation of the future is something only human beings can achieve. Bernstein argues that every significant act is guided by some model of a state of affairs that is to be brought about – a state of affairs that is not yet, that only 'may' be realized; and that the animal actively struggles towards this goal, using integrated programmes of behaviour aimed at the overcoming of obstacles which block the path. Thus intention, which becomes explicit in human consciousness, may be held to be implicit in all action – for Bernstein, even in the life of plants. Yet

clearly there are differences; and this brings me back to our starting-point. It seems to be true that the lower the point on the evolutionary scale, the more the 'models of future requirements' are fixed and determined, beyond the individual's control. Human beings are able to construct new ones in a rich diversity. Also, they can entertain more than one at a time. It is this which gives a choice of goals and no longer merely a choice of means in pursuit of a goal which imposes itself.

6. Bruner and Bruner, 'On Voluntary Action and Its Hierarchical Structure'; J. S. Watson, 'Perception of Contingency as a Determinant of Social Responsiveness' in E. Thomas (ed.), *The Origins of Social Responsiveness*, Hillsdale, New Jersey: Erlbaum, 1979. See also H. Papoušek, 'Individual Variability in Learned Responses in Human Infants' in R. J. Robinson (ed.), *Brain and Early Behaviour*, London: Academic Press, 1969. Researchers have reported different times of onset for this kind of behaviour, ranging from four or five weeks to three or four months. But DeCasper and his colleagues have even found that newborns can learn to regulate their sucking behaviour in the case where they are able to trigger on an audiotape the sound of their own mother's voice rather than the voice of a stranger. (See A. J. DeCasper and W. P. Fifer, 'Of Human Bonding: Newborns Prefer Their Mothers' Voices', *Science*, 1980, 208, pp. 1174–6.)

7. M. S. Mahler, F. Pine and A. Bergman, *The Psychological Birth of the Human Infant*, New York: Basic Books, 1967. See also L. Kaplan, *Oneness and Separateness*, London: Jonathan Cape, 1979.

8. J. Piaget, *The Child's Construction of Reality*, London: Routledge and Kegan Paul, 1955; first published in French 1936. Piaget claims that genuine search for hidden objects starts at between eight and ten months of age. Later researchers, however, have argued for earlier beginnings. For example, T. G. R. Bower and J. G. Wishart ('The Effects of Motor Skill on Object Permanence', *Cognition*, 1972, 1, pp. 28–35) found that, at as early as five months, infants who were plunged into darkness would reach for an object seen immediately before. This has been confirmed by B. Hood and P. Willatts ('Reaching in the Dark to an Object's Remembered Position: Evidence for Object Permanence in Five-month-old Infants', *British Journal of Developmental Psychology*, 1986, 4, pp. 57–65).

9. E. von Holst and H. Mittelstaedt, 'Das Reafferenzprinzip', *Naturwis-*

senschaften, 1950, 37, pp. 464–76. See also R. Held and A. Hein, 'Movement-produced Stimulation in the Development of Visually Guided Behaviour', *Journal of Comparative and Physiological Psychology*, 1963, 56, pp. 872–6. The quotation is from C. N. Johnson, 'Theory of Mind and the Structure of Conscious Experience' in J. W. Astington, P. L. Harris and D. R. Olson (eds.), *Developing Theories of Mind*, Cambridge: Cambridge University Press, 1988, p. 50.

10. P. J. Kellman, H. Gleitman and E. S. Spelke, 'Object and Observer Motion in the Perception of Objects by Infants', *Journal of Experimental Psychology: Human Perception and Performance*, 1987, 13, pp. 586–93. The evidence from their work may be interpreted as indicating that babies perceive *enduring* objects from the start; but whether or not this interpretation is accepted, the findings certainly seem to show that babies do not lack the distinction between self and 'something other'.

11. A. N. Meltzoff and W. Borton, 'Intermodal Matching by Human Neonates', *Nature*, 1979, 282, pp. 403–4.

12. Gibson, 'Exploratory Behavior in the Development of Perceiving, Acting and the Acquiring of Knowledge'. See also G. Butterworth, 'Events and Encounters in Infant Perception' in A. Slater and G. Bremner (eds.), *Infant Development*, Hove and London: Erlbaum, 1989. Butterworth, having reviewed the evidence, concludes that 'event perception is not a modality-specific process, rather it occurs by gathering of information from many sensory channels each attesting to the same external reality'. There is an interesting paper by Meltzoff and Moore on the importance for social development of the way in which supramodal representation makes possible the early imitation of others' facial expressions. (A. N. Meltzoff and M. K. Moore, 'Cognitive Foundations and Social Functions of Imitation and Intermodal Representation in Infancy' in J. Mehler and R. Fox (eds.), *Neonate Cognition: Beyond the Blooming Buzzing Confusion*, Hillsdale, New Jersey: Erlbaum, 1985.)

13. D. N. Stern, *The Interpersonal World of the Infant*, New York: Basic Books, 1985.

14. B. Dodd, 'Lip-reading in Infants: Attention to Speech Presented in and out of Synchrony', *Cognitive Psychology*, 1979, 11, pp. 478–84.

15. Confirmatory evidence that early speech perception is supramodal

278 Notes for pp. 40–44

has come from work by MacKain *et al.* (K. MacKain, M. Studdert-Kennedy, S. Spieker and D. N. Stern, 'Infant Intermodal Speech Perception is a Left-hemisphere Function', *Science*, 1983, 219, pp. 1347–9) and by P. Kuhl and A. N. Meltzoff ('The Bimodal Perception of Speech in Infancy', *Science*, 1982, 218, pp. 1138–41).

16. L. Murray, 'The Sensitivities and Expressive Capacities of Young Infants in Communication with Their Mothers', unpublished doctoral dissertation, University of Edinburgh, 1980. See also L. Murray and C. Trevarthen, 'The Infant's Role in Mother-Infant Communication', *Journal of Child Language*, 1986, 13, pp. 15–29. Other work bearing on the nature and consequences of disturbed interactions between infants and their mothers is discussed briefly but very clearly by T. Field in *Infancy*, Cambridge, Massachusetts: Harvard University Press, 1990 (a volume in the Developing Child Series, edited by Jerome Bruner and Michael Cole). Tiffany Field makes the point that it is possible for mothers to try too vigorously and intrusively to draw responses from their children, just as it is possible for them to give their babies insufficient attention and stimulation. Ideally the mother is highly sensitive to the signals that the infant gives. But depressed or anxious mothers, for instance, may not manage this easily. Field reports some attempts that have been made to improve disturbed interactions by teaching the mothers to do better, but she comments that, although these have seemed effective, it is hard to be sure that the gains carry over into everyday behaviour.

17. See, for instance, C. Trevarthen, 'Emotions in Infancy: Regulators of Contact and Relationships with Persons' in K. Scherer and P. Ekman (eds.), *Approaches to Emotion*, Hillsdale, New Jersey: Erlbaum, 1984.

18. It is clear that all activity of any kind occurs in the present – the 'here and now'. Nevertheless, the locus of concern may be the past or the future – or it may not lie in space-time at all.

19. The systematic combining of these modes in a single enterprise is the mark of science as it developed in the seventeenth and eighteenth centuries. This topic is further discussed in chapter 10. Notice that the use of two or more modes in deliberate combination, each having its assigned role, is a very different matter from the intermingling of components (emotion, thought, etc.) within a

mode. It is also different from the unplanned – indeed often involuntary – switching back and forth between modes that goes on very frequently in everyday living.

20. J. Galsworthy, *The Silver Spoon*, London: Heinemann, 1926, p. 65.
21. Stern, *The Interpersonal World of the Infant*.
22. Kaplan, *Oneness and Separateness*, pp. 135–6.
23. M. Konner, *The Tangled Wing: Biological Constraints on the Human Spirit*, Harmondsworth, Middlesex: Penguin, 1984, p. 223. (First published 1982.)
24. J. E. LeDoux, 'Cognitive-emotional Interactions in the Brain', *Cognition and Emotion*, 1989, 3, pp. 267–89.
25. Stern, *The Interpersonal World of the Infant*, p. 67.

4   The Onset of the Line Mode: Remembered Past and Possible Future

1. J. S. Bruner, 'The Growth and Structure of Skills' in K. J. Connolly (ed.), *Motor Skills in Infancy*, New York: Academic Press, 1971.
2. See Mary Warnock (*Memory*, London and Boston: Faber and Faber, 1987) for an interesting discussion of Proust's views on the distinction between this kind of memory, usually evoked by taste, smell, or some unexpected sensation, and the deliberate recall of 'pictures' of the past. Proust thought that the involuntary memories could sometimes produce extraordinary joy by enabling one to recapture the full rich truth of some past experience and thus, by joining past and present, allow us to know briefly what it is to escape into timelessness. On the other hand, he believed, as Warnock puts it: 'Nothing that we deliberately think up has this inbuilt guarantee of truth.'
3. See, for instance, A. E. Milewski and E. R. Siqueland, 'Discrimination of Colour and Pattern Novelty in One-month Infants', *Journal of Experimental Child Psychology*, 1975, 19, pp. 122–36. See also J. F. Fagan, 'Infants' Delayed Recognition, Memory and Forgetting', *Journal of Experimental Child Psychology*, 1973, 16, pp. 424–50.
4. A. J. DeCasper and M. J. Spence, 'Prenatal Maternal Speech Influences Newborns' Perception of Speech Sounds', *Infant Behavior and Development*, 1986, 9, pp. 133–50.
5. R. S. Lockhart, 'What Do Infants Remember?' in M. Moscovitch, (ed.), *Infant Memory*, New York: Plenum Press, 1984.

6. J. Kagan, R. B. Kearsley and P. R. Zelazo, *Infancy: Its Place in Human Development*, Cambridge, Massachusetts, and London: Harvard University Press, 1978.
7. L. Kaplan, *Oneness and Separateness*, London: Jonathan Cape, 1979.
8. H. R. Schaffer, A. Greenwood and M. H. Parry, 'The Onset of Wariness', *Child Development*, 1972, 43, pp. 165–75.
9. E. M. Cummings and E. L. Bjork, 'Perseveration and Search on a Five-choice Visible Displacement Hiding Task', *Journal of Genetic Psychology*, 1983, 142, pp. 283–91.
10. D. L. Schacter and M. Moscovitch, 'Infants, Amnesics and Dissociable Memory Systems' in Moscovitch (ed.), *Infant Memory*.
11. For instance, an adult subject who has learned a word list may or may not specifically remember a word from the list but, even if there is no conscious memory, it can still be shown that words on the list are later treated differently in certain respects from other words that were not on it.
12. K. Nelson, 'The Transition from Infant to Child Memory' in Moscovitch (ed.), *Infant Memory*. Katherine Nelson believes that the future is differentiated before the past. She bases this largely on the appearance of past and future markers in Emily's speech. I do not think that this is by any means conclusive.
13. Judy Dunn cites an example of reference to a specific earlier event by a child aged twenty-one months. (See J. Dunn, 'Understanding Feelings: The Early Stages' in J. S. Bruner and H. Haste (eds.), *Making Sense: The Child's Construction of the World*, London and New York: Methuen, 1987.) At breakfast on the same day the child and the mother had had a dispute about eating. The child started the later conversation, as follows:

CHILD: Eat my Weetabix. Eat my Weetabix. Crying.
MOTHER: Crying, weren't you? We had quite a battle. 'One more mouthful, Michael.' And what did you do? You spat it out!
[*Child pretends to cry.*]

The advent of talk about past events is further considered in chapter 7.
14. See, for instance, C. Trevarthen, 'Interpersonal Abilities of Infants as Generators for Transmission of Language and Culture' in A. Oliverio and M. Zapella (eds.), *The Behaviour of Human Infants*, London and

New York: Plenum Press, 1983; and 'Emotions in Infancy: Regulators of Contact and Relationships with Persons' in K. Scherer and P. Ekman (eds.), *Approaches to Emotion*, Hillsdale, New Jersey: Erlbaum, 1984.

15. J. Dunn, *The Beginnings of Social Understanding*, Oxford: Basil Blackwell, 1988.

5 'Pretend Play' and Conceptual Choice

1. A. Leslie, 'Some Implications of Pretense for Mechanisms underlying the Child's Theory of Mind' in J. W. Astington, P. L. Harris and D. R. Olson (eds.), *Developing Theories of Mind*, Cambridge: Cambridge University Press, 1988; and 'Pretense and Representation: The Origins of "Theory of Mind"', *Psychological Review*, 1987, 94, pp. 412–26.

2. J. Piaget, *Play, Dreams and Imitation in Childhood*, London: Routledge and Kegan Paul, 1951; first published in French 1946. The extracts quoted here are from pp. 96, 93 and 100 respectively.

3. James Russell would call this the onset of an 'asymmetry condition' between an object and the ways it can be represented. See his chapter 'Making Judgements about Thoughts and Things' (in Astington, Harris and Olson (eds.), *Developing Theories of Mind*) for some arguments about the ways in which we come to realize the distinction between the mental and the physical.

4. On teasing, see, for instance, V. Reddy, 'Playing with Others' Expectations: Teasing and Mucking About in the First Year' in A. Whiten (ed.), *Natural Theories of Mind*, Oxford: Basil Blackwell, 1991. On smiles of mastery, see, for instance, H. Papoušek, 'Individual Variability in Learned Responses in Human Infants' in R. J. Robinson (ed.), *Brain and Early Behaviour*, London: Academic Press, 1969.

5. Personal communication from Julie's mother. ('Julie', 'Rob' and 'Philip' are not the children's real names in this instance.)

6. M. Manning, J. Heron and T. Marshall, 'Styles of Hostility and Social Interactions at Nursery, at School and at Home' in L. A. Hersov and M. Berger (eds.), *Aggression and Anti-social Behaviour in Childhood and Adolescence*, London: Pergamon, 1978.

7. W. Shakespeare, *The Merchant of Venice*, I. iii. 43–4.

8. This is again a personal communication from the parent. And again the child's name has been changed.

6   *Two Varieties of the Construct Mode*

1. The quotation is from 'Some Causes of Bias in Expert Opinion', *The Psychologist*, 1989, 2, pp. 112–14. For a fuller discussion see J. St. B. T. Evans, *Bias in Human Reasoning: Causes and Consequences*, Brighton: Erlbaum, 1989.

2. See, for instance, M. Lewis and J. Brooks-Gunn, *Social Cognition and the Acquisition of Self*, New York: Plenum Press, 1979; and J. Kagan, *The Second Year of Life: The Emergence of Self-awareness*, Cambridge: Harvard University Press, Massachusetts, 1981. Colwyn Trevarthen shows, however, that interest in looking at mirrors has its origins very much earlier, at around the age of four or five months. He reports that, increasingly thereafter, babies 'seek the mirror, or rather its image of them, to appreciate their own enjoyment of play' and may make 'silly faces' at it. See C. Trevarthen, 'Signs before Speech' in T. A. Sebeok and J. Umiker-Sebeok (eds.), *The Semiotic Web, 1989*, Berlin: Mouton de Gruyter, 1990.

3. On the early use of personal pronouns, see Eve V. Clark, 'From Gesture to Word: On the Natural History of Deixis in Language Acquisition' in J. S. Bruner and A. Garton (eds.), *Human Growth and Development: Wolfson College Lectures, 1976*, Oxford: Clarendon Press, 1978.

4. G. B. Matthews, *Philosophy and the Young Child*, Cambridge, Massachusetts: Harvard University Press, 1980. See also M. Lipman and A. M. Sharp (eds.), *Growing Up with Philosophy*, Philadelphia: Temple University Press, 1978. Matthew Lipman has been influential in setting up the Institute for the Advancement of Philosophy for Children, which has come to be active in many countries throughout the world.

5. B. Tizard and M. Hughes, *Young Children Learning: Talking and Thinking at Home and at School*, London: Fontana, 1984.

6. P. L. Berger and T. Luckmann, *The Social Construction of Reality*, London: Allen Lane The Penguin Press, 1967, p. 154. (First published 1966.)

7. The conversation, quoted on p. 44, between a mother and child about why the moon could not be seen in the sky also shows that children have 'minds of their own' at quite an early age. I am indebted to Jess Reid for this example.

8. M. Hughes, *Children and Number*, Oxford: Basil Blackwell, 1986, p. 47.

9. A. N. Whitehead, *Science and the Modern World*, London: Free Association Books, 1985, p. 26. (First published 1926.)

10. Not, at least, until we see it in writing.

11. J. D. Bransford and N. S. McCarrell, 'A Sketch of a Cognitive Approach to Comprehension: Some Thoughts about Understanding What It Means to Comprehend' in W. B. Weimer and D. S. Palermo (eds.), *Cognition and the Symbolic Processes*, Hillsdale, New Jersey: Erlbaum, 1975.

12. In chapter 4 we discussed the kind of memory that Katherine Nelson emphasizes – memory built out of repeated experiences of similar kinds of event and yielding what we have to call *knowledge*. (See K. Nelson, 'The Transition from Infant to Child Memory' in M. Moscovitch (ed.), *Infant Memory*, New York: Plenum Press, 1984.) Nelson uses the term 'script' to refer to knowledge that tells us what to expect to happen in specific kinds of context. She also speaks of 'generalized event structures' and refers to these as the 'building blocks' of cognitive development. (See K. Nelson and J. Greundel, 'Generalized Event Representations: Basic Building Blocks of Cognitive Development' in M. E. Lamb and A. L. Brown (eds.), *Advances in Developmental Psychology*, Hillsdale, New Jersey: Erlbaum, 1981, vol. 1.) Daniel Stern thinks that, from early infancy onwards, personal interactions are generalized and turned into knowledge in the same kind of way. (See D. N. Stern, *The Interpersonal World of the Infant*, New York: Basic Books, 1985.)

13. This may not be true of some severely abnormal people, such as those who are autistic. Alan Leslie argues cogently for a connection between the disabilities of autistic children and a basic deficit in the capacity for pretence which leads to difficulties in understanding what is going on in the minds of others.

14. M. Hughes, 'Egocentrism in Pre-school Children', Edinburgh: Edinburgh University, unpublished doctoral dissertation, 1975. Also see M. Hughes and M. Donaldson, 'The Use of Hiding Games for Studying the Co-ordination of Viewpoints', *Educational Review*, 1979, 31, pp. 133–40.

15. S. Collie, 'Making Human Sense: A Study of Pre-schoolers' Ability to Hide Successfully in Sense and Nonsense Situations', unpublished MA dissertation, University of Edinburgh.

16. M. G. Dias and P. L. Harris, 'The Influence of the Imagination on Reasoning by Young Children', *British Journal of Developmental Psychology*, 1990, 8, pp. 305–18.

17. J. W. Astington, P. L. Harris and D. R. Olson (eds.), *Developing Theories of Mind*, Cambridge: Cambridge University Press, 1988. This book is an authoritative collection of articles by leading researchers. The editors' introduction provides an excellent overview of the available evidence and of competing interpretations.

18. However, there is also a great deal of further evidence. See, for instance, J. Flavell, S. G. Shipstead and K. Croft, 'Young Children's Knowledge about Visual Perception: Hiding Objects from Others', *Child Development*, 1978, 49, pp. 1208–11; M. Scaife and J. S. Bruner, 'The Capacity for Joint Visual Attention in the Infant', *Nature*, 1975, 253(5489), pp. 265–6; H. Borke, 'Piaget's Mountains Revisited: Changes in the Egocentric Landscape', *Developmental Psychology*, 1975, 11, pp. 240–43; and M. Schatz and R. Gelman, 'The Development of Communication Skills: Modifications of the Speech of Young Children as a Function of Listener', *Monographs of the Society for Research in Child Development*, 1973, p. 152.

19. H. Wimmer, G. J. Hogrefe and J. Perner, 'Children's Understanding of Informational Access as a Source of Knowledge', *Child Development*, 1988, 59, pp. 386–96.

20. H. Wimmer, G. J. Hogrefe and B. Sodian, 'A Second Stage in Children's Conception of Mental Life: Understanding Informational Accesses as Origins of Knowledge and Belief' in Astington, Harris and Olson (eds.), *Developing Theories of Mind*.

21. C. Pratt and P. Bryant, 'Young Children Understand that Looking Leads to Knowing (So Long as they are Looking into a Single Barrel)', *Child Development*, 1990, 61, pp. 973–82. It should be added that Wimmer has more recently modified his claims on the basis of his own further research. He still thinks that an understanding of the informational sources of belief has a part to play, but he no longer gives this a central role. He has come to agree with Perner that there is another fundamental prerequisite, namely the recognition that thoughts represent the world and may do so truly or falsely. Both Wimmer and Perner now hold that this realization ordinarily develops around the age of four. See H. Wimmer and M. Hartl, 'Against the Cartesian View on Mind: Young Children's Difficulty with Own False

Beliefs', *British Journal of Developmental Psychology*, 1991, 9, pp. 125–38.

22. K. Sullivan and E. Winner, 'When Three-year-olds Understand Ignorance, False Belief and Representational Change', *British Journal of Developmental Psychology*, 1991, 9, pp. 159–71.

7 *Language in Relation to the Modes*

1. D. Premack, 'Representational Capacity and Accessibility of Knowledge: The Case of Chimpanzees' in M. Piattelli-Palmarini (ed.), *Language and Learning: The Debate between Jean Piaget and Noam Chomsky*, Cambridge, Massachusetts: Harvard University Press, 1980.

2. Records collected in the Nursery School of the Department of Psychology, University of Edinburgh. Most of the children were not the sons or daughters of staff or students but came from a range of different backgrounds.

3. P. Hubley and C. Trevarthen, 'Sharing a Task in Infancy', *New Directions for Child Development*, 1979, 4, pp. 57–80.

4. E. Bates, *The Emergence of Symbols: Cognition and Communication in Infancy*, New York: Academic Press, 1979.

5. ibid., p. 35. Bates is here quoting a paper by E. Bates, L. Camaioni and V. Volterra, 'The Acquisition of Performatives prior to Speech', *Merrill-Palmer Quarterly*, 1975, 21, pp. 205–26. The example is of interest for showing how a child may persist until success is achieved. And commonly enough all does not go smoothly. The adult may sometimes not respond and may sometimes try to help but not understand. Golinkoff found that many of the communicative attempts made by pre-verbal children were not initially successful, so that the children had to give up or try again. Frequently they tried again, producing what Golinkoff calls 'repairs', that is, repeating or amplifying the first signal or using another in its place. There were three children in Golinkoff's study, aged between twelve and fifteen months when the research began. See R. M. Golinkoff, '"I Beg Your Pardon?": The Pre-verbal Negotiation of Failed Messages', *Journal of Child Language*, 1986, 13, pp. 455–76.

6. Sue Palmer, personal communication.

7. S. H. Foster, 'Learning Discourse Topic Management in the Preschool Years', *Journal of Child Language*, 1986, 13, pp. 231–50.

8. See, for instance, P. J. Miller and L. L. Sperry ('Early Talk about the

Past: The Origins of Conversational Stories of Personal Experience',
Journal of Child Language, 1988, 15, pp. 293–315) who studied five
working-class children between the ages of two years and two years
six months. It is of special interest that most references made by these
children to the past were about distressing events, especially those
involving physical harm. The quotations from Beth are thus shown
to be quite typical. See also J. Sachs, 'Talking about the There and
Then: The Emergence of Displaced Reference in Parent–Child
Discourse' in K. E. Nelson (ed.) *Children's Language*, Hillsdale, New
Jersey: Erlbaum, 1983, vol. 4. Sachs studied one child, Naomi, in
some detail between the ages of seventeen and thirty-six months. By
the end of this time, though Naomi 'could engage in conversations
about shared past experiences and non-present objects with no
support from the immediate situation' she still 'always returned to
comments about the immediate situation' very quickly. A. Preece
('The Range of Narrative Forms Conversationally Produced by
Young Children', *Journal of Child Language*, 1987, 14, pp. 353–73)
gives evidence of narrative language between the ages of five and
seven, by which time very considerable competence has developed.

9. D. I. Slobin and C. A. Welsh, 'Elicited Imitation as a Research Tool in
Developmental Psycholinguistics' in C. A. Ferguson and D. I. Slobin
(eds.), *Studies of Child Language Development*, New York: Holt,
Rinehart and Winston, 1973. Also see L. Bloom, 'Talking, Understanding and Thinking' in R. L. Schiefelbusch and L. L. Lloyd
(eds.), *Language Perspectives – Acquisition, Retardation and Intervention*,
New York: Macmillan, 1974.

10. N. Bernstein, *The Co-ordination and Regulation of Movements*, Oxford:
Pergamon, 1967.

11. A recent book by David McNeill (*Psycholinguistics: A New Approach*,
New York: Harper and Row, 1987) shows how important gestural
cues are in speech. David Olson has written a number of important
papers in which he discusses the differences between oral and written
language and considers the implications of these differences, both for
the development of individual minds and for the rise of modern
society. See, for instance, D. R. Olson, 'From Utterance to Text:
The Bias of Language in Speech and Writing', *Harvard Educational
Review*, 1977, 47, pp. 257–81; 'The Cognitive Consequences of
Literacy', *Canadian Psychology*, 1986, 27, pp. 109–21; and 'Mind and

Media: The Epistemic Function of Literacy', *Journal of Communication*, 1988, 38, pp. 27–35.

12. A. Bridges, 'Comprehension in Context', in M. Beveridge (ed.), *Children Thinking Through Language*, London: Edward Arnold, 1982.

13. Harris and Kavanaugh provide a very interesting example of spoken language that can be said to be produced 'to order' and that depends on construction of an imagined context, though with powerful support given. They report research in which two-year-olds watched while the experimenter pretended that a naughty teddy bear was pouring milk over a toy horse. The children were asked what they had seen. They described certain outcomes of the pretence – for instance, that the horse was wet – rather than the actual state of affairs. P. L. Harris and R. D. Kavanaugh, 'Young Children's Understanding of Pretense', *Society for Research in Child Development Monographs*, in press.

Later on, in school, children are frequently asked to write 'to order' on a theme that calls for the construction of an imagined context. Indeed, this has traditionally been very common and properly regarded as important. There has been a tendency of late to criticize it and to argue that children should write *only* when the writing will serve some clear purpose of their own. But this policy stems from a failure to recognize the importance of encouraging progress into the third and fourth modes. It is unfortunate that the policy is often presented as if it were particularly enlightened when it can, in fact, be quite limiting and stultifying if it is strictly adhered to. For further discussion of this topic see chapter 15.

14. L. A. French and K. Nelson, *Young Children's Knowledge of Relational Terms*, New York: Springer-Verlag, 1985.

15. M. L. Donaldson, *Children's Explanations: A Psycholinguistic Study*, Cambridge: Cambridge University Press, 1986; and 'Children's Comprehension and Production of Causal Connectives', paper presented at the Fourth International Congress for the Study of Child Language, University of Lund, Sweden, 1987. See also H. F. Emerson, 'Children's Comprehension of "Because" in Reversible and Non-reversible Sentences', *Journal of Child Language*, 1979, 6, pp. 279–300.

16. Shirley Brice Heath (*Ways with Words: Language, Life and Work in Communities and Classrooms*, Cambridge: Cambridge University

Press, 1983) reports an interesting study of early language socializa-
tion in three different communities and concludes that children who
are consistently encouraged at home to notice language and to 'talk
about talk' are the most likely to learn successfully when they go to
school. For a brief account see S. B. Heath, 'A Lot of Talk about
Nothing', *Language Arts*, 1983, 60, pp. 999–1007.

17. M. Hughes, *Children and Number: Difficulties in Learning Mathematics*,
Oxford: Basil Blackwell, 1986.

18. J. Piaget, *The Child's Conception of Number*, London: Routledge and
Kegan Paul, 1952. (First published in French 1941.)

19. J. McGarrigle, R. Grieve and M. Hughes, 'Interpreting Inclusion: A
Contribution to the Study of the Child's Cognitive and Linguistic
Development', *Journal of Experimental Child Psychology*, 1978, 25,
pp. 1528–50. See also T. Trabasso, A. Isen, P. Dolecki, A. McLanahan
and C. Riley, 'How Do Children Solve Class Inclusion Problems?'
in R. Siegler (ed.), *Children's Thinking: What Develops?*, Hillsdale,
New Jersey: Erlbaum, 1978.

20. B. Inhelder, H. Sinclair and M. Bovet, *Apprentissage et structures de la
connaissance*, Paris: Presses Universitaires de France, 1974. It does
seem to be the case that the inclusion relation presents problems to
young children over and above those raised by the comparison of
non-overlapping sets. See P. Josse, *Classes ou collections?*, Paris: Edi-
tions du Centre National de la Recherche Scientifique, 1984. See
also L. K. Miller and M. D. Barg, 'Comparison of Exclusive versus
Inclusive Classes by Young Children', *Child Development*, 1982, 53,
pp. 560–67.

The literature on the subject of class inclusion – and other tests
regarded by Piaget as criterial in deciding when a child's thought
becomes 'operational' – is now vast. A useful collection of papers
giving an overview of the main issues – and some idea of the
complexity of the arguments – was published in the *British Journal of
Psychology* in 1982 (vol. 73, pp. 157–311). My own paper in that
volume, entitled 'Conservation: What is the Question?', provides a
more complete statement of my views on the value of the Piagetian
tasks (as evidence for his theory and for other purposes) than it
would be appropriate to present here. However, I want to add a
further comment which relates to an article by Leslie Smith in the
same volume.

Smith argues strongly that the theoretical question which interested Piaget was whether children understand the *necessity* of the conclusion that a class contains more members than one of its subclasses (given, of course, that the other subclasses are not null, or empty). And he is certainly right: the understanding of necessary truth is what Piaget was trying to assess in the tasks he devised to study class inclusion, conservation and the like. Smith is also correct in saying that critics of Piaget sometimes lose sight of this, especially if they have read his works only in translation, where the emphasis on necessity may be weakened or lost.

The advent of the understanding of necessary truth greatly interests me also. What I doubt is whether the Piagetian tasks provide unambiguous evidence about it. Smith challenges the significance of James McGarrigle's finding that children make similar errors in cases where an inclusion relation exists and in cases where it does not. But his argument at this point puzzles me. He says:

To give a correct answer to a between-class question, such as 'Are there more white cows or more horses?', a child must understand that the class of horses is an including class in relation to its two subclasses. Thus, although the invited comparison does not embody an inclusion relation, it is a condition of the child's being able to quantify the membership of one of the classes that he should be able to understand inclusion.

What Smith appears to be saying here is that a child who does not understand the necessity of the relationships involved in the inclusion of subclasses within a superordinate class will not be able to count the superordinate class on its own. But I know of nothing to support such a claim; and Piaget himself does not make it. He does not argue that a pre-operational child cannot think of – or count – a whole class that has obvious subclasses. What is held to be impossible is to consider class and subclasses *simultaneously*. Where a child is dealing with a necklace of wooden beads, some of which are brown, then, to quote Piaget: 'If he is merely counting the wooden beads, he includes the brown ones, but if it is a question of counting first the brown ones and then the wooden ones, he . . . does not understand that the first set forms part of the second.' That is, counting a super-ordinate class – the horses, in McGarrigle's study – is not in itself held to be a problem. So Smith's argument does not seem to me to be any kind of defence of Piaget's position against McGarrigle's findings.

The question of how to know when a child understands the necessity of a correct conclusion remains. A number of years ago, working with children aged between four and seven, I spent some time trying to devise ways of determining this, but always I came up against the same difficulty: how to establish the difference between a child's conviction that she is making a correct judgement ('I am certain that . . .') and her conviction that something has to be the case ('It is certainly (i.e. necessarily) true that . . .'). Having failed to resolve this difficulty to my own satisfaction, I did not publish any account of the work.

The distinction between these two kinds of certainty is related to Frege's distinction between the content of a judgement and the justification of a judgement, or as Brian Rotman, following Frege, expresses it, between 'the psychological reasons for thinking X to be the case' and 'the logical arguments needed to justify X'. Rotman believes that Piaget was himself confused about this distinction and that the confusion vitiates his account of the nature of mathematical thought. See G. Frege, *The Foundations of Arithmetic*, Oxford: Basil Blackwell, 1950 (first published 1884), p. vi; and B. Rotman, *Jean Piaget: Psychologist of the Real*, Hassocks, Sussex: Harvester Press, 1977, p. 155.

21. This is equally true of statements. Alison Macrae ('Meaning Relations in Language Development', unpublished doctoral dissertation, University of Edinburgh, 1976) noted that there can be more or less natural descriptions. For instance, it is more natural to say: 'The flowers are on the table' than to say: 'The table is under the flowers.'

22. M. Donaldson and J. McGarrigle, 'Some Clues to the Nature of Semantic Development', *Journal of Child Language*, 1974, 1, pp. 185–194. James McGarrigle and I also studied the children's interpretations of quantifiers such as 'all'. For later relevant work see, for instance, N. H. Freeman and K. Schreiner, 'Complementary Error Patterns in Collective and Individuating Judgements: Their Semantic Basis in Six-year-olds', *British Journal of Developmental Psychology*, 1988, 6, pp. 341–50. See also M. Donaldson and P. Lloyd, 'Sentences and Situations: Children's Judgements of Match and Mismatch' in F. Bresson (ed.), *Problèmes actuels en psycholinguistique*, Paris: Centre National de la Recherche Scientifique, 1974.

23. For half the children, the two conditions were reversed. That is, first they saw the cars plus garages. Then the garages were taken away.
24. For more detailed discussion see M. Donaldson, *Children's Minds*, London: Fontana, 1978.
25. It turns out that the development of this attitude to language is a sophisticated achievement. See G. Bonitatibus, 'What is Said and What is Meant in Referential Communication' in J. W. Astington, P. L. Harris and D. R. Olson (eds.), *Developing Theories of Mind*, Cambridge: Cambridge University Press, 1988.
26. There arises also the question of helping children to think *about* language rather than *with* language. The importance in learning to read of having words for talking about language and knowing how to use them has been emphasized by a number of scholars. See, for instance, J. F. Reid, 'Learning to Think about Reading', *Educational Research*, 1966, 9, pp. 56–62; and D. R. Olson, ' "See! Jumping!" Some Oral Antecedents of Literacy', in H. Goelman, A. Oberg and F. Smith (eds.), *Awakening to Literacy*, London: Heinemann Educational, 1984.

8   The Intellectual Transcendent Mode

1. M. Hughes, *Children and Number: Difficulties in Learning Mathematics*, Oxford: Basil Blackwell, 1986. The work quoted was done with Miranda Jones and is reported in full in M. Jones, 'Children's Written Representations of Number and Arithmetical Operations', unpublished MA dissertation, University of Edinburgh, 1981.
2. It is evident, as Martin Hughes says, that such a child is completely unprepared to learn algebra. It should be added, in fairness, that the child who made this particular recommendation was only seven. But the study by M. Behr, S. Erlwanger and E. Nichols ('How Children View the Equals Sign', *Mathematics Teaching*, 1980, 92, pp. 13–15) covered the age range six to twelve. Similar results were obtained by A. Stallard ('Children's Understanding of Written Arithmetical Symbolism', unpublished MA dissertation, University of Edinburgh, 1982) whose subjects included children as old as ten years six months attending a middle-class school.
3. He also suggests some specific games that can be very valuable in helping young children to make a good start.
4. J. S. Bruner and H. J. Kenney, 'Representation and Mathematics

Learning', *Monographs of the Society for Research in Child Development* 1965, 30(1), pp. 50–59.

5. A. N. Whitehead, *Science and the Modern World*, London: Free Association Books, 1985, p. 33. (First published 1926.)

6. G. T. Kneebone, *Mathematical Logic and the Foundations of Mathematics*, London: Von Nostrand, 1963.

7. J. Piaget and B. Inhelder, *La Genèse de l'idée de hasard chez l'enfant*, Paris: Presses Universitaires de France, 1951.

8. Sometimes pairs containing two cards of the same colour (both blue, or whatever) were allowed, and sometimes not.

9. I have based the account in this paragraph on my own unpublished observations; but these agree quite closely with what Piaget and Inhelder found. They report that they encouraged children aged eight or more to '*trouver un truc pour être sûrs de les avoir tous*' (that is, 'find a way of being sure you've got them all'). I gave no such explicit encouragement, but the children I studied usually made the attempt just the same.

10. J. Piaget, 'Intellectual Evolution from Adolescence to Adulthood', *Human Development*, 1972, 15, pp. 1–12.

11. J. Piaget, *Logic and Psychology*, Manchester: Manchester University Press, 1953. A good, balanced critical account of Piaget's work can be found in Margaret Boden's book, *Piaget* (London: Fontana, 1979). See especially the chapter entitled 'Logic in Action'. Brian Rotman's book, *Jean Piaget: Psychologist of the Real* (Hassocks, Sussex: Harvester Press, 1977) is also valuable, particularly on the subject of Piaget's ideas about the nature of mathematics.

12. This work was not published.

13. G. Frege, in the introduction to *The Foundations of Arithmetic* (Oxford: Basil Blackwell, 1986), comments on what E. Schröder had called 'the Axiom of Symbolic Stability' which, as Schröder put it, 'guarantees us that throughout all our arguments and deductions the symbols remain constant in our memory – or preferably on paper'. Frege is scornful of Schröder for treating this as an axiom and so 'confusing the grounds of proof with the mental and physical conditions to be satisfied if the proof is to be given'. And Frege is, of course, right. But as a psychological precondition for logical thought the importance of 'symbolic stability' can scarcely be overstated.

14. Hughes, *Children and Number*. See also Jones, 'Children's Written Representations of Numbers and Arithmetical Operations'.

PART TWO: RANGE AND BALANCE IN MATURITY

9 *The Intellectual and Value-sensing Modes: A Look at History*

1. Notice, however, that as individuals most of us are not perhaps so very one-sided after all – and this not because of advanced emotional development but rather because of lack of intellectual development. It is only a few among the world's millions whose intellects are the source of the power that science and technology have given us. The rest of us have derivative power only. But the derivative power is still immense. A president or prime minister whose intellect may be by no means outstanding is not thereby prevented from ordering the pressing of a button. And, to take a more ordinary case, no great intellectual skill is needed to pass a driving test; yet, having done this, we are all legally endowed with the power to command a fast and highly dangerous machine. Thus the intellectual development of a few has brought vast derived power to the many. So even if the few were just as advanced emotionally and morally as intellectually, the problem of achieving wise and good use of our derived power would remain.

2. One way to redress the balance would of course be to curtail the intellect and return to some 'simpler' way of life. But this is a very negative solution; and it is also very unlikely to be resorted to, unless massive ecological disaster forces it upon us.

3. J. M. Plumley, 'The Cosmology of Ancient Egypt' in C. Blacker and M. Loewe (eds.), *Ancient Cosmologies*, London: George Allen & Unwin, 1975.

4. H. R. Ellis Davidson, 'Scandinavian Cosmology', in Blacker and Loewe (eds.), *Ancient Cosmologies*, p. 182.

5. Plumley, 'The Cosmology of Ancient Egypt'.

6. Phrase contained in a Spell from the Coffin Texts, quoted by Plumley, 'The Cosmology of Ancient Egypt', p. 25.

7. A. N. Whitehead, *Science and the Modern World*, London: Free Association Books, 1985, p. 238.

8. E. Cassirer, *An Essay on Man*, New Haven: Yale University Press, 1944, p. 92.

9. C. B. Boyer, *A History of Mathematics*, Princeton, New Jersey:

Princeton University Press, 1985, p. 5. (First published 1968.) G. G. Joseph (*The Crest of the Peacock: Non-European Roots of Mathematics,* London and New York: I. B. Tauris, 1991) gives some helpfully detailed accounts of how the Egyptian calculations were performed. He thinks there is evidence for the beginnings of what he calls 'rhetorical algebra' – that is, algebraic reasoning without a symbolic system in which to express it. Boyer also thinks that some problems in the Ahmes papyrus are algebraic since, in effect, they entail the solving of linear equations containing unknowns. And both writers comment on another document, the Moscow papyrus, dating from around 1890 BC. This shows that the Egyptians had a method for calculating the volume of a truncated pyramid which is in accordance with the modern formula. How the ancient method was arrived at is not known.

10. O. Neugebauer, *The Exact Sciences in Antiquity,* New York: Harper and Row, 1962. (First published 1952.)
11. ibid., p. 42.
12. K. Ward, *Images of Eternity,* London: Darton, Longman and Todd, 1987.
13. An introduction to the life and work of St John of the Cross, together with a selection of his writings, is provided in K. Kavanaugh's book, *John of the Cross,* London: SPCK, 1987, and Mahwah, New Jersey: Paulist Press, 1987. The following quotations are from pp. 106–7, 119, 139 and 100 respectively.
14. Ward, *Images of Eternity,* p. 3.
15. P. Toynbee, *Towards the Holy Spirit,* London: SCM Press, 1982.
16. Ward, *Images of Eternity,* p. 130.
17. ibid., p. 115.
18. ibid., pp. 157 and 159.

10   *The Modes and the Advent of Science*

1. Deanna Kuhn and her colleagues have studied the development of scientific thinking in children growing up in the United States today. Their main conclusion is that competence depends on making a clear, conscious distinction between theory and evidence before bringing these into an 'ideal co-ordination' which will cut out the distorting effects of personal preferences or preconceptions. Kuhn *et*

al. conclude from their studies that the skills involved 'are weak among children below adolescence . . . show some development from middle childhood to adulthood, but . . . remain at far less than an optimal level of development even among adults'. See D. Kuhn, E. Amsel and M. O'Loughlin, *The Development of Scientific Thinking Skills*, New York: Academic Press, 1988. The quotation above is from p. 220.

2. A. C. Crombie, *Augustine to Galileo: Science in the Middle Ages* (2 vols.), Harmondsworth: Peregrine Books, 1969, vol. 1, p. 40. (First published 1952.)

3. K. Ward, *Images of Eternity*, London: Darton, Longman and Todd, 1987. One could question whether the dominant mode between the ninth and twelfth centuries was not the core rather than the value-sensing construct mode. I think the latter, for the reason that attempts at explanation seem to have been positively disapproved of, as the next part of the discussion will show.

4. Crombie, *Augustine to Galileo*, vol. 1, p. 35.

5. Quoted by Brian Stock from a twelfth-century commentary on Virgil's *Aeneid*, attributed to Bernard Silvester. See B. Stock, *Myth and Science in the Twelfth Century*, Princeton, New Jersey: Princeton University Press, 1972.

6. Quoted by Stock, *Myth and Science in the Twelfth Century*, p. 41. However, this quotation may be somewhat unrepresentative of Erigena's thinking, even if not of the prevailing modes. His great philosophical work *De divisione naturae* makes it clear that he himself was far from regarding rational thought about the nature of things as an activity to be shunned.

7. C. B. Boyer, *A History of Mathematics*, Princeton, New Jersey: Princeton University Press, 1985, p. 276. The Italian mathematician referred to is Leonardo of Pisa, known as 'Fibonacci'. It should be added to Boyer's assessment that Leonardo had had a Muslim teacher and was much influenced by Arabic methods. He is best remembered for introducing Hindu-Arabic numerals to Europe.

8. Given the variety of the sources, the commonly held idea that the Renaissance in Europe was attributable solely or even mainly to a recovery of ancient Hellenistic learning is clearly flawed. George Gheverghese Joseph stresses this in his book *The Crest of the Peacock: Non-European Roots of Mathematics* (London and New York: I. B.

Tauris, 1991). He regards the belief as just one manifestation of a general tendency on the part of Europeans to take a Eurocentric view of the history of knowledge. He is no doubt right that such a tendency is strong and needs to be corrected. However, many other scholars have acknowledged the contributions of such great civilizations as Babylon, India and China. For instance, Charles Everitt, writing on algebra in the *Encyclopaedia Britannica*, eleventh edition, recognizes the extraordinarily impressive contributions of Indian mathematicians to the development of that subject. And Boyer (in *A History of Mathematics*) certainly does not overlook it or play it down.

9. Crombie, *Augustine to Galileo*, vol. 1, p. 45. Another translator was Michael Scot, a Scottish philosopher and mathematician who, having learned Arabic at Toledo, moved to the court of Frederick II in Sicily, where he was encouraged to translate Aristotle. He also wrote works of his own on astrology and alchemy, and he acquired a great reputation as a magician.

10. I am indebted to Rosalind Mitchison for pointing out that, during this period, considerable intellectual effort was devoted to matters of law, especially canon law.

11. A. N. Whitehead, *Science and the Modern World*, London: Free Association Books, 1985. (First published 1926.)

12. H. Butterfield, *The Origins of Modern Science*, London: G. Bell and Sons Ltd, 1949, p. 7. Butterfield also points out that there arose in both Oxford and Paris, in the fourteenth century, schools of thought critical of aspects of Aristotelian doctrine. And the scholars involved, men like Jean Buridan and Nicholas of Oresme, anticipated some of the profound changes in conceptions of the universe that were to come. Nicholas went so far as to propose that the universe might prove to be a kind of clock, started by God, certainly, but then left to run on its own. This tradition of thought lasted continuously into the sixteenth century and is believed to have been known to Leonardo and to Galileo. See also Crombie, *Augustine to Galileo*, vol. 2, for a discussion of Buridan's theory of *impetus* which, in important respects, anticipated Newton.

Paul Harris suggests that Aristotle's ideas on motion were so hard to overcome because of the way they fit with human sense. (Harris, personal communication.) On the subject of human sense, see p. 95 of this book.

13. Butterfield, *The Origins of Modern Science*, p. 5. Notice that what Galileo imagined was, strictly speaking, impossible in the physical world; yet it proved highly relevant to an understanding of the physical world.

14. Crombie, *Augustine to Galileo*, vol. 1.

15. F. A. Yates, *The Rosicrucian Enlightenment*, Frogmore, St Albans: Paladin, 1975. (First published 1972.) See also M. L. Righini Bonelli and W. R. Shea, *Reason, Experiment and Mysticism in the Scientific Revolution*, London and Basingstoke: The Macmillan Press, 1975.

16. Yates, *The Rosicrucian Enlightenment*, p. 75.

17. ibid., pp. 139 *et seq.*

18. G. Naudé, *Apologie pour les grands hommes soupçonnés de Magie*, Paris, 1625.

19. A. Baillet, *La vie de Monsieur Descartes*, Paris, 1691.

20. Yates, *The Rosicrucian Enlightenment*, p. 231.

21. ibid., p. 232.

22. Crombie, *Augustine to Galileo*, vol. 1, p. 181.

23. See, for instance, R. S. Westfall, 'Alchemy in Newton's Career', and P. Casini, 'Newton, a Sceptical Alchemist' in Righini Bonelli and Shea, *Reason, Experiment and Mysticism*.

24. Butterfield, *The Origins of Modern Science*, p. 24.

25. P. Rossi, 'Hermeticism, Rationality and the Scientific Revolution' in Righini Bonelli and Shea, *Reason, Experiment and Mysticism*.

26. ibid., p. 270.

27. B. Barnes, *About Science*, Oxford: Basil Blackwell, 1985.

28. ibid., p. 66.

11   *The Advanced Modes after the Enlightenment*

1. P. Redondi, *Galileo Heretic*, London: Allen Lane The Penguin Press, 1988. (First published 1983.)

2. R. Descartes, *A Discourse on Method*, London: J. M. Dent and Sons, 1912, Part IV, first paragraph. (First published in French 1637.)

3. Bernard Le Bovier de Fontenelle, *The Utility of Mathematics*, Paris: Collected Works, 1790. The quotation can be found in H. Butterfield, *The Origins of Modern Science*, London: G. Bell and Sons Ltd, 1949, p. 173.

4. I do not intend to imply that this was something new upon the face of the earth. Secular societies have existed in other times and places,

as Mary Douglas (*Natural Symbols*, London: Barrie and Rockliff, 1970) reminds us. It was, however, a new thing in Europe by contrast with many preceding centuries.

5. W. Law quoted in Bullett, *The English Mystics*, London: Michael Joseph, 1950, p. 150.

6. W. Wordsworth, *The Prelude* (1850 edition). The following quotations are from Book II, ll. 302–37, 357–8, 213–15 and 216–19 respectively.

7. C. Darwin, *Autobiography*, London: Collins, 1958. This and the following quotations are from pp. 91, 138 and 139 respectively.

8. Adrian Desmond (*The Politics of Evolution*, Chicago and London: University of Chicago Press, 1989) gives a detailed and very interesting account of how, in the decades preceding the publication of *The Origin of Species*, evolutionary notions were strongly advocated by some and strongly resisted and condemned by others on grounds that had more to do with political conviction than with scientific evidence or rational argument. He reaches the overall conclusion that 'the rival biological doctrines in the thirties were integrated into long-term commercial and political strategies, either to gain or to hold on to privileges' (p. 23). This amounts, in my terminology, to saying that the line and core construct modes were very actively influencing the debates, as they often do since they are hard to keep out. Darwin was so aware of all this that he delayed publication of his major work for about twenty years, for fear, Desmond tells us, of social ostracism.

9. A. Alland, Jr, *Human Nature: Darwin's View*, New York: Columbia University Press, 1985. This quotation is cited on p. 173. The following two are on p. 176 and p. 181 respectively. All are taken from *The Descent of Man*.

10. Paul Davies gives clear expositions of some of the mind-boggling conclusions that have emerged. See especially his chapter 'The Nature of Reality' in *Other Worlds: Space, Superspace and the Quantum Universe*, London: Sphere Books, 1982. (First published 1980.)

11. A. N. Whitehead, *Science and the Modern World*, London: Free Association Books, 1985, p. 45. (First published 1926.)

12. Davies, *Other Worlds*, p. 122.

13. E. Wigner, *Symmetries and Reflections*, Bloomington and London: Indiana University Press, 1967, p. 172. Eugene Wigner goes on to

wonder how materialism could so long have been accepted by most scientists. His answer is that it is probably an emotional necessity to exalt the problem to which one wants to devote a lifetime. He adds the suggestion that 'the principal problem is no longer the fight with the adversity of nature but the difficulty of understanding ourselves if we want to survive' (p. 177).

14. Davies, *Other Worlds*, p. 75.

15. One disturbing development is the trend towards a kind of popularizing which extends the new ideas far beyond the sub-atomic sphere without making any serious attempt to justify this. For example, T. N. Griffiths writes: 'Character only gains meaning within a non-mechanistic, quantum world-view, where the world is indeterminate for us.' And again: 'The changed environment will unpredictably stress and reveal new personality features of other people and oneself (Heisenberg's *Uncertainty Principle*), about which we then make further choices between possible responses.' Both quotations are from: 'Teaching and Learning the Trinity-Unity Model of Man', *Holistic Medicine*, 1988, 3, pp. 175–84.

16. W. James, *The Varieties of Religious Experience*, London and New York: Longmans, Green, 1902. This quotation is on p. 491. The quotations that follow are on pp. 519, 122, 47 and 110 respectively.

12   *Dealing with Emotions: Some Western Ways*

1. See S. Vosniadou and W. F. Brewer ('Theories of Knowledge Restructuring in Development', *Review of Educational Research*, 1987, 57, pp. 51–67) for a discussion of the kind of learning or 'knowledge restructuring' that the second example entails. For a general discussion see also S. Carey, *Conceptual Change in Childhood*, Cambridge, Massachusetts: MIT Press, 1985.

2. L. Thomas, *The Medusa and the Snail*, Harmondsworth, Middlesex: Penguin Books, 1981, p. 15. (First published 1979.)

3. M. Milner, *A Life of One's Own*, London: Virago, p. 71, 1986. (First published 1934 under the pseudonym Joanna Field.)

4. Freud thought that the defence mechanisms came into operation between the ages of two and five. See S. Freud, *Analysis Terminable and Interminable* (1937, Standard Edition, vol. 23), where he recognizes the price we often have to pay for such benefits as they bring.

5. See H. S. Hughes, *Consciousness and Society*, Brighton: Harvester Press, 1979. (First published 1958.) The statements about virtue and vice, and about nature and history are from Taine. Hughes does not give their exact source.

6. T. Nagel, *The View from Nowhere*, Oxford: Oxford University Press, 1986, p. 191.

7. B. Bettelheim, *The Informed Heart*, Harmondsworth, Middlesex: Penguin Books, 1986, p. 16. (First published 1960.)

8. M. Milner, *A Life of One's Own*.

9. W. James, 'What is an Emotion?', *Mind*, 1884, 9, pp. 188–205.

10. For the most part, the source of these events is the experience of the individual. Some, however, may have deeper origins in the experience of the race – or at any rate there are those who believe this to be possible.

11. It is not relevant here to engage in any argument about the respective roles of language and images in thought. This has been much debated, but I think no one would doubt that both have a part to play. We may all in certain circumstances – usually when powerful emotion is being experienced – find ourselves 'at a loss for words' or 'speechless'. Or we may find ourselves saying things we later regret. So sometimes emotion controls speech, not vice versa. Also people may 'dry up' because of shyness, lack of confidence or the stress of some unusually daunting moment. And everyone makes a certain number of 'slips of the tongue': that is, we all occasionally say things that we did not mean to say. As Freud realized, these slips are not just amusing. They may reveal aspects of our minds that we are reluctant to acknow; and they certainly indicate that conscious control of speech has its limits. K. W. Fischer, P. R. Shaver and P. Carnochan ('How Emotions Develop and How they Organize Development', *Cognition and Emotion*, 1990, 4, pp. 81–127) give examples, based on empirical research, of what they call 'adult scripts' for emotions. A script consists of 'antecedents', 'responses' and 'self-control procedures'. The emotions exemplified in the paper are anger and joy. In the case of anger, self-control procedures consist in attempts to 'suppress or hide the anger or redefine or remove the situation so that anger is no longer called for'. In the case of joy, self-control is said to be, typically, not a salient issue. However, 'suppression of joy in the interest of decorum or avoidance

of envy is possible'. So on this account American adults today are not so very different in their practices from Jane Austen's heroine. Fischer and his colleagues do remark, though, that their subjects may have made 'subtle or less frequent attempts to control positive emotions' in spite of not mentioning them.

12. Thoughts 'occur to us', they 'pop up', they even 'strike us'. Also thoughts wander. Yet, on the other hand, they can become obsessive, hard to get rid of. They can plague and haunt us. A relevant question is: where do questions come from? For questions are crucial for the conduct of thought, from its most primitive to its most sophisticated forms. Sometimes we know very well where our questions come from, but not always. For example, programmes of scientific inquiry, however systematic and elaborate, depend for their success on the generation of good hypotheses, which are in effect questions. But we have no formula that we can apply to produce them. Hypotheses are among the kinds of idea that tend to 'strike' us. They come to us; but we do not know from where, or how to guarantee their coming. Some minds seem to generate them more readily than do others. We call such minds creative.

13. R. Ainsworth, 'The Ugsome Thing' in *Ten Tales of Shellover*, London: André Deutsch, 1963.

14. To get a sense of how cognitive therapy is used see, for instance, J. Scott, J. M. G. Williams and A. T. Beck (eds.), *Cognitive Therapy in Clinical Practice: An Illustrative Casebook*, London and New York: Routledge, 1989.

15. Freud, *Analysis Terminable and Interminable*, 1937, Standard Edition, vol. 23, pp. 229–30.

16. Bettelheim, *The Informed Heart*. The following quotations are from pp. 18, 19, 20 and 27 respectively.

17. G. Jantzen, *Julian of Norwich: Mystic and Theologian*, London: SPCK, 1987, pp. 184–5.

13 *Dealing with Emotions: Some Buddhist Ways*

1. Or if *winning* the game comes to be what matters, they remain in the line mode, but with a change of focus that takes them away temporarily from the source of distress. Paul Harris's book *Children and Emotion* (Oxford: Basil Blackwell, 1989) gives a very valuable

account of when and how children come to develop insight into their own emotions, the emotions of other people and the ways in which these interact. He surveys research on these topics, much of it carried out by himself and his colleagues.

2. R. Kipling, 'How the Camel Got his Hump', in *Just So Stories*, Harmondsworth: Penguin Books, 1988. (First published 1902.)

3. A. N. Whitehead, *Science and the Modern World*, London: Free Association Books, 1985. (First published 1926.)

4. The research by Harris and Guz is reported in Harris, op. *Children and Emotion*, pp. 161 *et seq.*, and in P. L. Harris and G. R. Guz, 'Models of Emotion', unpublished paper, Department of Experimental Psychology, University of Oxford, 1986.

5. H. Saddhatissa, in *The Buddha's Way* (London: George Allen & Unwin, 1971) gives an unusually accessible yet authoritative account of what the Buddha was 'getting at', to use his own words. I am very grateful to Tom Thorpe for giving me a copy of this work when he learned that I was trying to understand Buddhism.

6. A. W. Watts, *The Way of Zen*, Harmondsworth, Middlesex: Penguin Books, 1962. (First published 1957.)

7. The word often translated as 'suffering' is *dukkha*. Paul Griffiths (*On Being Mindless*, La Salle, Illinois: Open Court, 1986, p. 150) suggests 'unsatisfactoriness' as a good alternative. Watts (*The Way of Zen*, p. 66) proposes 'frustration' or 'sourness'.

8. Watts, *The Way of Zen*, p. 67.

9. Griffiths, *On Being Mindless*, p. 15.

10. W. L. King, *Theravāda Meditation: The Buddhist Transformation of Yoga*, University Park and London: The Pennsylvania State University Press, 1980, p. 92.

11. E. Fromm, 'Psychoanalysis and Zen Buddhism' in D. T. Suzuki, E. Fromm and R. de Martino, *Zen Buddhism and Psychoanalysis*, New York: Grove Press, 1963. Fromm argues that the links between Zen and psychoanalysis are close; but while he has much to say that is interesting and relevant I do not think he makes a wholly convincing case.

12. King, *Theravāda Meditation*, p. 34.

13. This is by no means unique to Buddhism. The Christian who renounces the world often does so in the hope of attaining paradise.

14. King, *Theravāda Meditation*.

15. The term 'mindlessness' is the one Paul Griffiths uses. The difficulty of distinguishing the ultimate state of cessation from death was recognized and much debated. Only those automatic processes essential to life seem to have been held to continue. Once again it is necessary to stress the diversity of Buddhism. There are many branches within which 'cessation' is not, and has not ever been, regarded as a supreme achievement. Nevertheless, Buddhaghosa, who after all was chosen by Keith Ward as an impeccably orthodox figure, did evaluate it very highly. So we cannot dismiss it as a mere aberration.

16. I am conscious that I am far from having done justice to the great Mahayana school of Northern India. But this book is not a history of Buddhism and, so far as I understand the matter, I have omitted nothing that is of great relevance here.

17. D. T. Suzuki 'Lectures on Zen Buddhism' in Suzuki, Fromm and De Martino, *Zen Buddhism and Psychoanalysis*, pp. 52 *et seq.*

18. Sokei-an Sasaki is cited by Watts in *The Way of Zen*, on p. 141. Watts gives the original source as 'The Transcendental World', *Zen Notes*, vol. 1, no. 5, New York: First Zen Institute of America, 1954.

19. Watts, *The Way of Zen*, p. 140.

20. See Saddhatissa (*The Buddha's Way*, p. 83.) who speaks of the Theravādin Buddhist belief that, after four experiences of Path consciousness, total liberation or enlightenment is achieved. I do not know why four is thought to be the critical number.

21. Suzuki, 'Lectures on Zen Buddhism', p. 16.

22. Watts, *The Way of Zen*. This quotation and the following ones are from chapters 2 and 3.

23. King, *Theravāda Meditation*, p. 122.

24. Saddhatissa (*The Buddha's Way*, pp. 54 *et seq.*) warns us, however, against being too 'objective' and 'mechanistic'. We are to be detached and observant, watching calmly, 'allowing each state to come and go unimpeded'. But we are not to think of ourselves as things. And we are to include ourselves in the '*metta*' – or loving-kindness – that we are to feel for all others under the code of Buddhist morality.

25. K. Ward, *Images of Eternity*, London: Darton, Longman and Todd, 1987. The quotations are from pp. 63 and 69.

26. In some traditions, meditation is distinguished from 'concentration'

and/or 'contemplation'; but in much modern usage these distinctions are not drawn and the word 'meditation' has come to serve as a general term. I shall use it in this way.

27. E. Wood, *Yoga*, Harmondsworth, Middlesex: Penguin Books, 1962, p. 76.

28. Griffiths, *On Being Mindless*.

29. J. Blofeld, *The Tantric Mysticism of Tibet*, London: George Allen & Unwin, 1970.

30. For a survey of research on this subject, see M. A. West (ed.), *The Psychology of Meditation*, Oxford: Clarendon Press, 1987. According to Michael West – and other contributors – there is no solid body of evidence that meditation reduces the physiological effects of stress as it is often claimed to do, especially by those trying to 'sell' it today. West argues, however, that research has concentrated too much on looking for quick changes of this kind during meditation practice instead of asking about the possible occurrence of long-term changes in those who meditate regularly.

31. E. Pagels, *The Gnostic Gospels*, Harmondsworth, Middlesex: 1982, p. 141. (First published 1979.) Pagels offers an extended discussion of this aspect of Gnosticism. She believes that other branches of the early Christian Church felt greatly threatened by the placing of so much reliance on personal experience rather than on revealed authority. She also shows that the Nag Hammadi texts contain prescriptions for meditative practices as a technique for moving towards enlightenment.

32. Wood, *Yoga*, p. 225. Of course Buddhists do not give accounts of this kind of unitive experience in terms of 'self', even written with a capital letter. For them the ideal is loss of self. But the transcendent experience seems to be very similar, as far as one can tell.

33. Plato, *The Republic*, VII.

34. *The Republic*, VI, 510.

35. *The Republic*, VI, 508 and 509.

36. I. Murdoch, *The Sovereignty of Good over Other Concepts*, Cambridge: Cambridge University Press, 1967, pp. 21–2.

37. Plato, *Philebus*, 52c.

38. I. Murdoch, *The Fire and the Sun: Why Plato Banished the Artists*, Oxford: Clarendon Press, 1977.

14   *Value-sensing Experiences: Some Contemporary Evidence*

1. D. Hay, *Exploring Inner Space* (2nd edition), London and Oxford: Mowbray, 1987.
2. M. Laski, *Ecstasy: A Study of Some Secular and Religious Experiences*, London: The Cresset Press, 1961. Laski acknowledges her amateur status as a researcher but she has a very keen intellect and a scrupulous respect for empirical inquiry. Her study is valuable. However, it must be borne in mind that her subjects provided a very unrepresentative sample of the population.
3. D. Hay, 'The Bearing of Empirical Studies of Religious Experience on Education', *Research Papers in Education*, 1990, 5, pp. 3–28. The quotation is from p. 15. The example is drawn from the archives of the Alister Hardy Research Centre. These contain over 5,000 accounts sent in response to mass-media appeals.
4. G. Ahern, *Spiritual and Religious Experience in Modern Society: A Pilot Study*, Oxford: The Alister Hardy Research Centre, 1990. Ahern himself emphasizes that this is a pilot study. The sample is quite small and the results should be treated with caution.
5. As David Hay puts it: 'There seems to be a feeling that "society" in some way does not give permission for these experiences to be integrated into ordinary life.' And people are covertly angered by this, as Hay discovered. He asked for descriptions of the sorts of people who would never report such experiences and found what he calls 'a catalogue of extraordinary vehemence', attributing to the 'non-reporters' such qualities as apathy, conformity, insensitivity, superficiality, lack of imagination, cowardice, and so on. It was an interesting question to ask. Hay comments on the lack of any evidence for such judgements. He takes them to be a way of expressing rage against society for not letting us acknowledge our own inner perceptions. See Hay, *Exploring Inner Space*, pp. 164–5.
6. Hay also discusses the evidence of links with education, reviewing a number of studies which report a statistical connection. He points out, however, that there are confounding factors such as age and race; for black people are more likely than white to report religious experiences. See *Exploring Inner Space* and the paper quoted in note 3.
7. In *Exploring Inner Space* Hay remarks that most of his own subjects

could recall no specific external triggers, but that around half of them were distressed or ill at the time the experience occurred. So there is disparity here.

8. Laski, *Ecstasy*. This extract and the following one are from pp. 176 and 183.

9. Ahern, *Spiritual and Religious Experience in Modern Society*.

10. M. Milner, *A Life of One's Own*, London: Virago, 1986. (First published 1934 under the pseudonym Joanna Field.) The quotations I have used so far are from pp. 22, 28 and 31. Those that follow are from pp. 106 *et seq.*, 123, 150 and 186. Milner acknowledges a number of influences on her thinking in spite of her resolve to find her own way. She had been impressed at the start of her enterprise by the essays of Montaigne and particularly by his notion that the soul may turn out to be quite different from what one supposes it to be. (See the introduction to *Eternity's Sunrise*, her most recent book – London: Virago, 1987.) She also found Piaget's early work on children's thought and language helpful in the analysis of some of her own immature ways of thinking. And she was influenced by Freud's notion of the unconscious mind, though she did not use detailed Freudian theory at this stage. Later she trained as a psychoanalyst. In an epilogue to *A Life of One's Own*, Milner tries to use the Jungian notion of the 'masculine' and 'feminine' in each of us to explain some of her findings. However, in my own opinion this adds no valuable insights.

11. M. Milner, *An Experiment in Leisure*, London: Virago, 1986. (First published 1937 under the pseudonym Joanna Field.)

12. D. Freedberg (*The Power of Images: Studies in the History and Theory of Response*, Chicago and London: University of Chicago Press, 1989) has some very relevant things to say about the nature of our responses to images and about why we 'love art and hate it; we cherish it and are afraid . . .' (p. 388). He discusses all manner of public imagery, not just what we ordinarily call 'art'.

13. Milner, *An Experiment in Leisure*, p. 226.

15 *Other and Better Desires: Prospects for a Dual Enlightenment*

1. This is not to deny the existence of limitation due to handicap. For example, it is hard to see how even Beethoven could have composed

his music if he had been *born* deaf. Yet he wrote his finest works in spite of becoming deaf. Much handicap can be overcome.

2. A. N. Whitehead, *The Aims of Education*, London: Ernest Benn, 1936, p. 15. (First published 1932.)

3. M. Donaldson, *Children's Minds*, London: Fontana, 1978. For examples of the concept as I define it now, see the present volume, chapter 6.

4. See the essay entitled 'Common Sense as a Cultural System' in C. Geertz, *Local Knowledge*, New York: Basic Books, 1983. The quotation is from p. 85.

5. I do not mean to imply that deliberate learning is *absent* in the early years of life. The point is to notice how much is then achieved spontaneously.

6. In this connection a very difficult question arises about segregated schooling in any culture where subgroups coexist and are aware of their sharp distinctiveness. There can be no doubt that cultural distinctiveness yields part of one's sense of having a personal secure identity. The question, however, is whether minority distinctiveness should be fostered at school, or whether school should aim to unify a whole society. Debate about this often rages most fiercely in relation to the teaching of religion and of history. But even in the teaching of mathematics cultural influences are seldom absent. The intellectual construct mode involves the use of imagined contexts. And if there is insensitivity or carelessness on the teacher's part, then 'simple sums' can call up cultural images which might offend or undermine a child's sense of identity. This is certainly also true of story-books. And it is hard for the writers of these books to stay imaginatively aware of the extent of the problem. I am grateful to Alison Elliot for helpful discussion of this topic.

7. Piaget was convinced – correctly in my opinion – that it is a gross error to think of children as passive recipients of 'input', 'stimulation' or 'information'. This led him to emphasize the child's own construct-ive activities. He did not, in fact, completely ignore social influences, but very often he did not make much reference to them. Thus people in general, and educators in particular, have tended to draw from his work the conclusion – quite wrong in my opinion – that children in all the most important respects can find their own way.

8. J. S. Bruner, 'The Growth of Mind', *American Psychologist*, 1965, 20,

pp. 1007–17. I was privileged to take some part in the research which Bruner here describes.

9. A. N. Whitehead, *The Aims of Education*, p.19.

10. *Yani's Monkeys*, Beijing: Foreign Languages Press, 1984. The quotation is from the introduction entitled 'The Heart of a Child', by Huang Qingynn. (The pages are not numbered.)

11. ibid. The quotation is from the foreword by Jiang Feng.

12. J.-P. Sartre, *Words*, London: Hamish Hamilton, 1964, p. 30.

13. R. Kipling, *Kim*, London: Macmillan, 1930. (First published 1908.)

14. 'Satire on the Trades' in J. B. Pritchard (ed.), *Ancient Near-Eastern Texts*, Princeton, New Jersey: Princeton University Press, 1955.

15. I have discussed the topic of literacy and intellectual development in earlier publications; see *Children's Minds*; see also *Sense and Sensibility: Some Thoughts on the Teaching of Literacy* (Reading: University of Reading, Reading and Language Information Centre, 1989). It seems to me that *how* literacy is taught matters greatly – and that this is true from the start of the teaching.

16. H. S. Hughes, *Consciousness and Society: The Reorientation of European Social Thought, 1890–1930*, Brighton: Harvester Press, 1979, p. 27. (First published 1958.)

17. F. Braudel, *On History*, London: Weidenfeld and Nicolson, 1980, p. 215. (First published in French 1969.)

18. Braudel speaks of the importance of this too, mentioning Lucien Febvre's work, 'La Religion de Rabelais', in which Lebvre tried to work out 'what range of words, concepts, reasoning and sensibilities' Rabelais would have had available to him. Notice the inclusion of 'sensibilities'. See Braudel, *On History*, p. 208.

19. P. Rossi, 'Hermeticism, Rationality and the Scientific Revolution', in M. L. Righini Bonelli and W. R. Shea (eds.), *Reason, Experiment and Mysticism in the Scientific Revolution*, London: Macmillan, 1975, p. 272.

Index